A Strategic Digital Transformation for the Water Industry

Edited by

Oliver Grievson
Timothy Holloway
Bruce Johnson

Published by

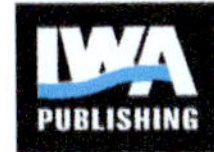

IWA Publishing
Republic - Export Building, 1st Floor
2 Clove Crescent, E14 2BE
London, United Kingdom

T: +44 (0)20 7654 5500
F: +44 (0)20 7654 5555
E: publications@iwap.co.uk
W: www.iwapublishing.com

First published 2022
© 2022 IWA Publishing

ISBN: (Paperback) 9781789063394
ISBN: (eBook) 9781789063400
DOI: 10.2166/9781789063400

DISCLAIMER

This book is a collection of the 9 white papers published in the IWA Digital Water Programme's White Paper Series.

The core content and message of the white papers remains the same. However, additional information has been added to maintain the topics up to date.

COPYRIGHT NOTICE CHAPTER 2

About the editors

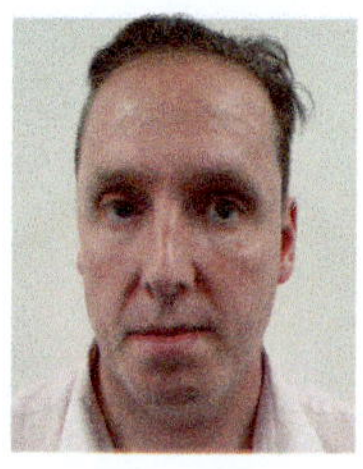

Oliver Grievson

Technical Lead – Water Industry at Z-Tech Control Systems Ltd
Chair of the IWA Digital Water Programme Steering Committee

Oliver Grievson is a wastewater instrumentation specialist and process engineer with over 25 years of experience working in the water industry in the UK and internationally. He is also an expert in the Digital Transformation of the Water Industry and is currently serving as the Chairman of the IWA Digital Water Programme.

He is a Chartered Engineer, Chartered Environmentalist, Chartered Scientist and Chartered Water & Environmental Manager as well as being a Fellow of several professional organisations. In his daily job for Z-Tech Control System, he acts as a Technical Lead helping numerous clients with wastewater monitoring ranging from the scoping of complex capital projects to acting as an Expert Witness in legal disputes. He also volunteers for a number of organisations in the field of wastewater and is keen on the development of the future generations of instrumentation engineers.

Timothy Holloway

Senior scientific officer at the University of Portsmouth
Senior management committee member for the IWA Modelling and Integrated Assessment (MIA) specialist group

Timothy Holloway is a wastewater technology and modelling engineer specialising in visualising the dynamic resilience of Water Resource Recovery Facilities (WRRF) in response to climate change. Before entering the water industry, Timothy worked as a project and design engineer for the civil and structural framing industry. With over 10 years of experience in the water industry, he started as a process design engineer and then Process Engineering Manager for a WRRF manufacturer, managing a team of engineers on bespoke international industrial and municipal wastewater projects. Then he became a Senior Scientific Officer as part of an innovation Hub collaboration between the University of Portsmouth and Southern Water. The project focused on testing various technologies and systems, primarily around the low P challenge. More recently, Timothy has been working collaboratively with Southern Water on an Ofwat Innovation fund project to develop intensive digital monitoring solutions.

Timothy is also a researcher for the Environmental Technology and Management Research Group at the University of Portsmouth and is involved in developing applied WRRF resilience methodologies. Currently, this involves combining various methods working toward autonomous solutions for the water industry in the UK. As a Chartered Engineer, Timothy mentors young engineers and scientists on their chartership journey. He has recently become a senior member of the IWA MIA specialist group and hopes to combine the benefits of modelling with digital systems.

Bruce R. Johnson

PE, BCEE, Technology Fellow at Jacobs
IWA Fellow

Bruce Johnson is a wastewater technology Senior Fellow with Jacobs and has led the process design for most of its wastewater-related alternative delivery projects, as well as being the global technology leader for wastewater process simulation. As part of his work within Jacobs, he has been responsible for developing and implementing a wide range of innovative treatment approaches, including advanced control approaches.

He has been doing wastewater treatment design for over 30 years, the last 27 of which has been with CH2M/Jacobs. He started out his career working in the technology group of Eimco (predecessor to Ovivo) where he worked on various biological and filtration technologies. With CH2M, he has worked in its industrial, design and water business groups where he has held the roles of Industrial Process Practice Leader and Wastewater Process global technology leader.

He has been active outside of Jacobs in WEF and IWA. Within WEF, he led the development of the Nutrient Removal MOP, vice chaired the committee that developed the Wastewater Simulation MOP and was a contributing author in MOP 8's suspended growth chapter. He also chaired the WEFTEC wastewater symposia committee and was a founding member and past chair of WEF's Modeling Experts Group of the Americas. Within IWA, Mr. Johnson helped develop the wastewater treatment modelling seminar series (WWTmod) and is a past scientific chair. He also was a founding member of the IWA Design and Operations Uncertainty Task Group and is currently participating in the Physical Chemical Modelling Task Group as well as an Industrial Liaison within the Modelling & Integrated Assessment management committee.

In his career, Bruce has published many papers on a wide range of wastewater treatment subjects and has been awarded four different patents based on his innovative approaches to wastewater treatment, design and control.

The International Water Association's role in the digital journey

The International Water Association (IWA) can influence and facilitate change as it is a place for water professionals from across the spectrum to make these changes happen, including conducting in-depth research and creating thought leadership in creative partnerships. In a complex dynamic changing world, the Association can build bridges between silos, linking outcomes across sectors, and raising awareness and urgency in the political arena. It is with this mindset and combined with leading-edge scientific breakthroughs, technological developments and creative mindsets that we can challenge the complexities of the water sector now and in the future.

The IWA's 5-year Strategic Plan recognises the need for innovation in the global water sector to address the challenges associated with global change pressures. IWA realises that transformation cannot take place on its own in a vacuum. The ideas for solutions to these challenges will be fomented and rigorously debated among the IWA's institutions and membership in an open, yet controlled space.

This book is a compilation of the knowledge shared and generated so far in the *IWA Digital Water Programme*. The Programme acts as a catalyst for innovation, knowledge, and best practice, and provides a platform to share experiences and promote leadership in transitioning to digital water solutions. By sharing experience on the drivers and pathways to digital transformation in the water industry, the programme is consolidating lessons and guidance for water utilities to start or continue to build their journey towards digitalisation. The ecosystem of IWA members across the water value chain including utilities, regulators, technology companies, software companies, researchers and academia will be at the forefront of embracing emerging technologies to solve urgent and costly issues around water service provision (operation, liability, customer services, etc.).

Through the Digital Water Programme, the IWA leverages its worldwide member expertise to guide a new generation of water and wastewater utilities on their digital journey towards the uptake of digital technologies and their integration into water services.

The IWA Digital Water Programme Steering Committee from left to right:
Row 1: Oliver Grievson (Chair), Hélène Hauduc, Marina Batalini de Macedo, Eunice Gnay Namirembe
Row 2: Biju George, Sheilla de Carvalho, Vladan Babovic, Rik Thijssen
Row 3: Randolf Waters, Jyoti Gautam, Frank Kizito, Zoran Kapelan
Row 4: Enrique Cabrera, Rachel Peletz, Yufeng Guo, Cecilia Wennberg
Row 5: Saba Daneshgar, Ann Piyamarn Sisomphon, Deepa Karthykeyan, Zach White, Stuart Hamilton

Contributors

Chapter 1

Randolf Waters *Senior Director of Global Partnerships & Branding, Xylem Inc.*
Will Sarni *CEO, Water Foundry*
Cassidy White *Consultant, Water Foundry*
Katharine Cross *Strategy and Partnerships Lead, Australian Water Partnership*
Raül Glotzbach *Scientific Researcher, KWR Water Research Institute*

Chapter 2

Oliver Grievson *Technical Lead, Z-Tech Control Systems Ltd.*
Daniel Aguado *Associate Professor, Institut Universitari d'Investigació d'Enginyeria de l'Aigua i Medi Ambient (IIAMA), Universitat Politècnica de València, València*
Frank Blumensaat *Senior Research Fellow, ETH Zurich/Eawag, Institute of Environmental Engineering, Chair of Urban Water Management Systems*
Juan Antonio Baeza *Full Professor, Department of Chemical, Biological and Environmental Engineering, Universitat Autònoma de Barcelona, Cerdanyola del Vallès*
Kris Villez *Sr. R&D Staff Member, Oak Ridge National Laboratory*
Maria Victoria Ruano *Associate Professor, Chemical Engineering Department, Universitat de València*
Oscar Samuelsson *Researcher, IVL Swedish Environmental Research Institute*
Queralt Plana *R&D Engineer, modelEAU, Université Laval*
Dragan Savic *Chief Executive Officer, KWR Water Research Institute*
Peter van Thienen *Senior Researcher & Chief Information Officer, KWR Water Research Institute*

Chapter 3

Bruce Johnson *Technology Fellow at Jacobs*
Borja Valverde- Pérez *Assistant Professor at Department of Environmental Engineering, Technical University of Denmark*
Christoffer Wärff *Industrial PhD student at RISE Research Institutes of Sweden / Lund University*
Douglas Lumley *DHI Sweden Ab*
Elena Torfs *Assistant Professor at Department of Data analysis and Modelling, Ghent University*
Ingmar Nopens *Full Professor, BIOMATH, Ghent University*
Lloyd Townley *Chief Scientist at Nanjing University Yixing Environmental Research Institute and Founder of Nanjing Smart Technology Development Co. Ltd*
Zoran Kapelan *Delft University of Technology, Department of Water Management*
Emma Weisbord *Digital Water Partnerships, Royal HaskoningDHV Digital*
Vladan Babovic *National University Singapore, Department of Civil and Environmental Engineering*
Timothy Holloway *Senior Scientific Officer at University of Portsmouth School of Civil Engineering and Surveying*

John Barry Williams *Professor of Environmental Technology at University of Portsmouth School of Civil Engineering and Surveying*
Djamila Ouelhadj *Professor of Operational Research at the University of Portsmouth Department of Mathematics*
Gong Yang *Process Capacity Manager at Southern Water Services Ltd*
Ben Martin *Lead Research Scientist at Thames Water*
Matthew Wade *Science Lead: Wastewater Research and Development, Environmental Monitoring for Health Protection, Data Analytics and Surveillance at the UK Health Security Agency*

Chapter 4

Alex Knezovich *Toilet Board Coalition*
Brian Moloney *Co-Founder and Managing Director of StormHarvester*
Jonathan Lavercombe *Chief Operating Officer at StormHarvester*
Jody Knight *Asset Technology Manager at Wessex Water*
Jacobo Illueca *Wastewater Specialist at Idrica*
Raimundo Seguí *Epidemiologist at Global Omnium*

Chapter 5

Hagimar von Ditfurth *Digital Ambassador for GIZ Uganda*
Emma Weisbord *Digital Water Partnerships, Royal HaskoningDHV Digital*
Thor Danielsen *Sector Expert – Water at the Embassy of Denmark in Poland*
Lloyd Fisher-Jeffes *Digital Water Partnerships, Royal HaskoningDHV Digital*
Ana Claudia Hafemann *Regulatory Analyst at AGIR*
Juljana Hima *Organization for Security and Cooperation in Europe*
Temple Chukwuemeka Oraeki *State Focal Person, Consultant at British Council*
Oliver Grievson *Technical Lead, Z-Tech Control Systems Ltd.*

Table of Contents

Forewords

The importance of the digital transformation to the global water industry

The pressing need to address the challenges of the global water sector continues to be the driving force behind the digital transformation. In addition to rapidly changing demographics, intensified and prolonged natural disasters, and ageing infrastructure, the sudden emergence of new public health threats has added pressure on utilities, regulators, and governments. In the face of the COVID-19 pandemic, utilities were forced to accelerate implementation of digital solutions to maintain communication with customers and staff, and to monitor and control essential assets. Despite the unfavourable nature of these events, they provided opportunities for us to observe the benefits of digitalisation.

From International Water Association's (IWA) viewpoint, we can see that the digital transformation is progressing at an impressive rate. So, to help catalyse innovation and smarter water management, we are pleased to share this selection of collective experiences and expert knowledge of our members to encourage you to internalise and envision what digitalisation may look like from your context.

This book is a collection of papers published under the IWA Digital Water Programme's White Paper series. At the time of writing, we have published nine such papers, authored by our diverse and expert members. The book gives insight into some of the best practices found in digitalisation, combining academic research and industrial applications, and presenting case studies of successful implementation. We want this book to be a guide to utilities, water managers, water stewards, and everyone interested in the digital transformation journey.

The IWA Digital Water Programme is underpinned by the work and experiences of utilities, technology and software companies, regulators, and researchers involved in the digital transformation of the water sector. IWA continues to maintain a space where practitioners, both accustomed and new to digitalisation, can share with, learn from, and inspire each other and the wider water sector. Hence, this publication fits directly into the Programme's and IWA's strategy and objectives.

We at IWA continue to lead the global discourse on digital water by leveraging the deep wells of proficiency within our membership. In this era, only by sharing our collective knowledge and promoting collaboration can we ensure the ongoing improvement of our sector. Doing so will improve efficiency in the sector, reduce our carbon footprint, and ultimately lead to improved outcomes in water management and sanitation for all.

Kalanithy Vairavamoorthy
Executive Director of the International Water Association

Academia's role in digital transformation

Digital transformation has a range of benefits to offer to the water industry and wider society, as clearly stated by the authors of this book. This transformation is a complex and lengthy process and can only be achieved by different stakeholders working together. This includes scientists and academics located in universities and various other research organisations.

One of the key roles that academia has to play in the digital transformation of water industry is in conducting fundamental research, i.e., developing new methods, technologies and knowledge that underpin this transformation. Indeed, developing state-of-the-art smart water solutions, both hardware (instrumentation such as various sensors mentioned in Chapter 2) and software (digital twins and tools based on artificial intelligence mentioned in Chapter 3) is not possible without the corresponding fundamental research, and that is what universities and research organisations do best. In addition to highly skilled and educated people, academia has access to state-of-the-art research infrastructure, such as various labs and high-power computing facilities, that is critical for the development of new smart water technologies. The fundamental research, of course, needs to be conducted in collaboration with other stakeholders such as technology development companies, industry regulators, various consultants and especially the end users (water utilities). This way research relevance and benefits can be ensured for all parties involved, including the members of general public, via improved public health (see Chapter 4 of this book) and otherwise.

Another equally important role that academia has to play, in the digital transformation of the water industry and society, is educational. Developing or using the new smart water technologies presented in Chapters 2-4 of this book requires a new generation of employees who understand the concepts of, and have skills in, data science, artificial intelligence, machine learning, advanced measurement, instrumentation techniques and so on, all in addition to more conventional water engineering knowledge. Among others, a new generation of the so-called 'hydroinformaticians,' i.e., people who work on the borderline of two disciplines, water engineering and information and communication technologies, will need to be educated. The need for these and similar profiles is likely to increase exponentially in the future, as indicated in the last chapter of this book.

Academia also has a significant role to play in overcoming various barriers to the adoption of new digital technologies, both in the water industry and the society. Several barriers have been identified and discussed in the first and last book chapters. If not addressed well, the barriers have the potential to slow down significantly the digital transformation process or even prevent some aspects of it. Academia can contribute to overcoming these by a number of different means, e.g. by providing rigorous scientific evidence and demonstrations that new methods and technologies work, by educating different stakeholders and the general public ensuring improved understanding and hence acceptance of new technologies, or by simply facilitating the dialogue between interested parties. Academia can also help by developing new, step-change technologies that are more user friendly and transparent, i.e., easier to understand and apply.

Digitalisation and related technologies have a great potential to improve everyday water management. The journey to a digital water future has started already but there is still a long way to go. Academia has multiple important roles to play in this journey. The key, however, is in all interested parties working closely together to co-create new technologies, solutions, and knowledge, improving the quality of our lives.

Zoran Kapelan
Digital Water Programme Steering Committee Member
Chair and Professor of Urban Water Infrastructure — University of TU Delft

Digitalisation in Practice

 A survey of utilities within both developed and emerging economies has revealed that nearly all utilities have begun the journey towards digital transformation. This transformation is inevitable as the industry faces risks from increasing water demand, water scarcity, declining water quality, ageing and underfunded infrastructure, climate change, rapid urbanisation and other global change pressures. As they strive to counter these pressures, utilities find themselves at varying levels of digital maturity. And as digital technologies grow exponentially in capabilities and decline exponentially in costs, agile and dynamic utilities have been quick to embrace the benefits of digitalisation, with many leapfrogging to adopt emergent technologies.

Nearly all digital solutions have a positive impact on reducing operational expenditure, increasing capital efficiency, improving the customer experience, enhancing the efficiency of the utility workforce and promoting the utility brand. These benefits notwithstanding, utilities face barriers to the uptake of digital solutions. In developing countries, resources to invest in digital technologies are sparse and compete in priority with investments in water infrastructure. Many legacy systems and assets cannot readily adapt to digitalisation requirements, necessitating costly investments in their retrofitting or replacement. The introduction of digital technologies may necessitate a paradigm shift from traditional workflows, requiring staff retooling and workforce change management. Cybersecurity, coupled with a perceived invasion of privacy, poses a challenge from the customer perspective. In some countries, lack of an enabling framework of relevant policies and regulations is a barrier to the adoption of digital technologies.

This book introduces the building blocks of 'Digital Water', describes its enabling technologies and ever-widening ecosystem of stakeholders, and presents practical case studies of digitalisation across various components of the utility value chain. It highlights the role played by digital technologies in the transformation from data to information and from information to insight and makes a case for the adoption and scaling of digitalisation by utilities in both developed and emerging economies, irrespective of which stage along the digital water adoption curve they currently are. As such, it is beneficial to utilities as a tool for peer learning within the digital water space, alongside IWA's Digital Water White Paper series and other related resources.

To be ready for a "digital water future", utilities must be enabled by policy and regulation to adopt innovative technologies and ways of working. Furthermore, a utility must decide which of its processes to prioritise for digitalisation. To this end, IWA's five-phase "digital water adoption curve", which is both an assessment tool and a roadmap to guide utilities in their digital adoption journey, is a useful resource. And as water and wastewater utilities become more digital, opportunities for establishment of collaborative ecosystems are enhanced, with improved ability to share critical data and insights with cities, government agencies, utilities within other sectors and the general public. Indeed, as posited in this book, utilities that have embraced digitalisation are uniquely positioned to act as a "springboard to a Smart City".

Frank Kizito

IWA Digital Water Programme Steering Committee Member
Decision Support Systems Manager — National Water and Sewerage Corporation
Water utility practitioner

Chapter 1

Defining Digital Water

1.1 Introduction

There have never been more reasons to feel optimistic about our water future. Innovations in digital solutions are unlocking the ability to deliver transformative outcomes to communities around the world, and businesses, utilities, and governments are on the cusp of taking needed actions to address the most pressing water challenges of affordability, scarcity, and resilience.

The 'Digital Water: Industry leaders chart the transformation journey' white paper was released in June 2019, serving as the launch of the International Water Association's Digital Water Programme. Based on interviews, surveys and inputs from nearly 50 utility executives and over 20 subject matter experts, the white paper provided an overview of the state of digital transformation in the water sector, the value potential for digital solutions, and the lessons learned from water and wastewater utilities. This book introduction expands the information provided in the white paper and sets the tone for the book.

From proactively managing aging infrastructure, to ensuring water quality from source to tap, to advancing water equity, and much more, the solutions to address these challenges are now helping global utilities navigate the complexities of driving improvements across the entire water cycle. There has never been a better time to be a utility professional.

The building blocks of digital water

The story of Cape Town, South Africa is one that has been told countless times as a worst case scenario – a countdown to 'Day Zero,' the day a city runs out of water. As Cape Town continues to struggle to address its water shortage, similar challenges are arising across the globe. Sao Paulo, Brazil faces challenges in water supply. Shenzhen, China faces challenges with untreated sewage discharges. Miami, United States, faces challenges with sea-level rise. Jakarta, Indonesia, faces challenges with groundwater depletion. The list goes on and on.

In the face of such extreme water challenges, water and wastewater utilities have been compelled to turn to new and innovative solutions: digital technologies.

Umgeni Water, a water utility in Durban, South Africa, has used digital technology to better manage its water resources and protect its customers from the same fate as Cape Town residents. Hydrologic models paired with monitoring devices have allowed the Durban water utility to optimise storage levels in dams and reservoirs. Likewise, the Las Vegas Valley Water District has harnessed digital technology to reduce non-revenue water, improving conservation and optimising water supply for customers. In Shenzhen, water quality monitoring sensors and hydraulic modelling systems implemented by Shenzhen Water Group have resulted in vast improvements to surface water quality. Moving forward, as global stressors continue to exacerbate current water challenges, digital solutions will be necessary for addressing the various problems utilities face to ensure adequate, reliable services to customers.

Digital technologies offer unlimited potential to transform the world's water systems, helping utilities become more resilient, innovative, and efficient, and in turn helping them build a stronger and more economically viable foundation for the future. Exploiting the value of data, automation, and artificial intelligence allows water utilities to extend water resources, reduce non-revenue water, expand infrastructure life cycles, provide the basis for financial security, and more.

The water sector's value chain links the environment and water resources to a utility, the utilities to their customers, and the customers back to their environment. From physical infrastructure to water quality to customer service and beyond, digital water can be integrated at every key point across the water cycle.

It is important to note that the value chain for water extends beyond the boundaries of a utility to include water sources (e.g., the watershed and other sources) and users (e.g., the public sector and industries). This is reflected in the *IWA Action Agenda for Basin-Connected Cities*, which outlines the rationale and provides a framework to harmonise urban, industrial, agricultural and ecological demands within a watershed.

Deploying these digital solutions across the expanded view of the value chain, and within the steps of the value chain, is

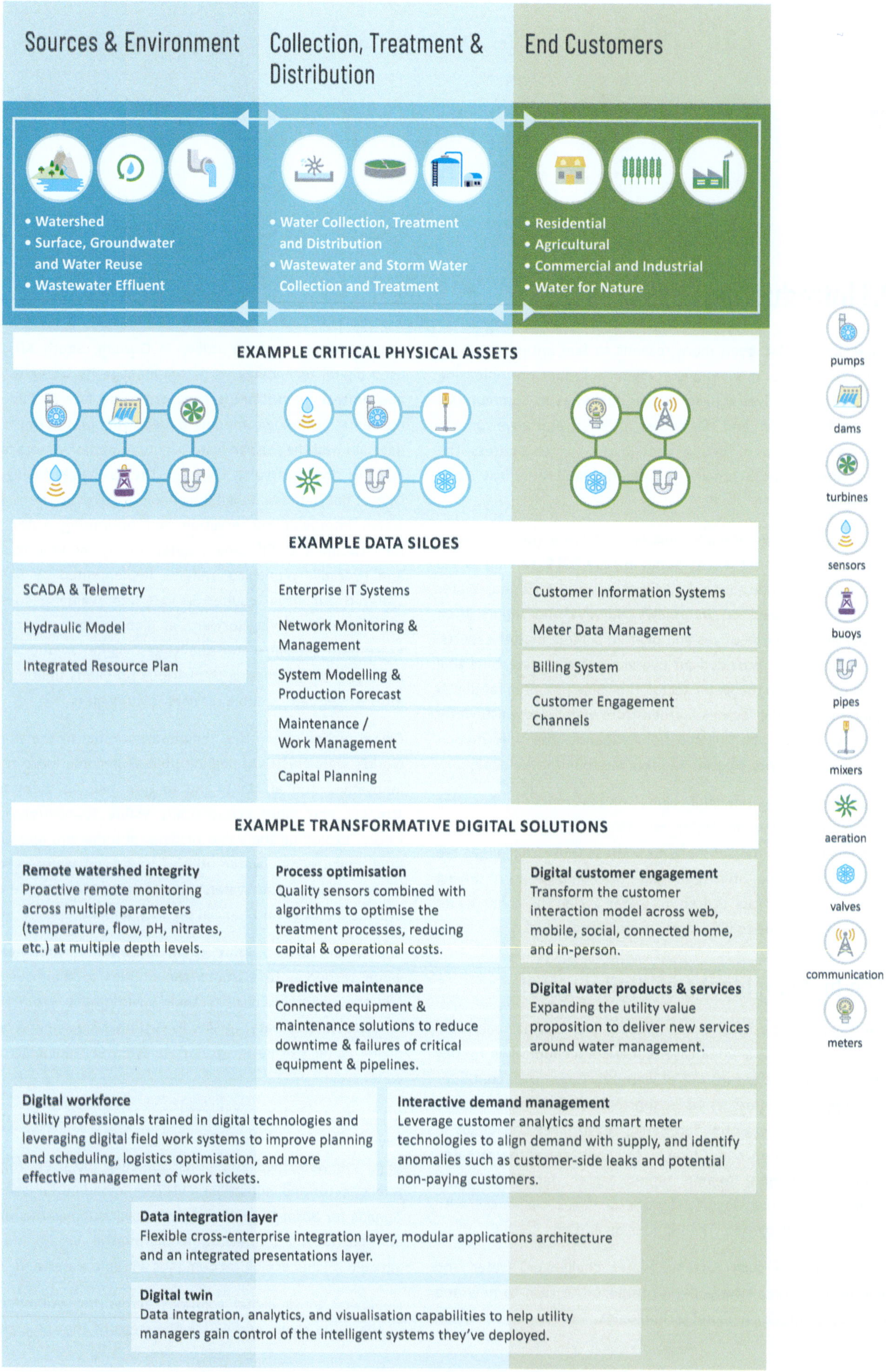

Figure 1.1 Technology Inputs and Solutions of a Digital Water and Wastewater Utility

easier said than done. Utilities are complex organisations, with multiple departments that each come with their own objectives, organisational layers, networks of physical assets, and siloes of data systems. Figure 1.1 further establishes the key components of the technology inputs and solutions within a water and wastewater utility organisation.

With increasing complexity in water systems and management, there is growing potential and need to adopt transformative digital solutions. For example, remote sensing and digital twin technologies provide connectivity between a utility and its diversified water supply. Various digital technologies then provide connectivity within a utility's operations. Customer service and customer analytics tools are then positioned to bridge the gap between a utility and its customer and initiatives, such as open data platforms and citizen science projects, can provide connectivity from the customer back to their water supply. These solutions all leverage the latest in enabling technologies: cloud, mobile, intelligent infrastructure, sensors, communication networks, and analytics and big data.

The enabling technologies of digital water

There are many digital solutions that are part of a utility, and even more technology providers and start-ups which are facilitating their design, installation, and operation. Many of these solutions leverage the latest innovations seen across industries: advanced sensors, data analytics, blockchain integration, and artificial intelligence. Given the accelerating rate of innovation in these foundational technologies, below is a snapshot of current exciting innovations .

SENSORS, MONITORING AND FORECASTING

Sensors, remote sensing, geographic information systems (GIS) technologies, and visualisation tools are becoming key elements for managing water resources at service area, watershed and regional scales. Remote sensing/imaging technologies, such as satellites and drones, used separately or together, provide data for mapping water resources, measuring water fluxes and utility asset management. Data from such technologies can better prepare water resource managers and utilities for incidents of heavy storm water flow (e.g., altering operations to prevent sewage overflow), indicate when conservation practices should be enacted during periods of drought, and ensure all treated water is delivered to customers. In addition, satellite data can be used to provide water quality data (e.g., turbidity, algal blooms, etc.) and hydrological forecasts, which, when used in conjunction with in situ measurements, allow utility operators to prepare for and react to water quality issues and other challenges.

New and existing sensors, both fixed and mobile, are being used to provide near real-time data on water quality, flows, pressures, and water levels, among other parameters. Sensors can be dispersed throughout systems to aid daily operations by optimising resource use (e.g., chemical use for water treatment), detect, diagnose, and proactively prevent detrimental events (e.g., pipe bursts, water discoloration events, sewer collapses/ blockages, etc.), and provide useful information for preventative maintenance and improved longer-term planning for water utilities (e.g., by helping to prioritise repairs and replacements for aging infrastructure). Likewise, sensors can provide evidence for pipe corrosion and alert home owners and utilities when water quality standards are not being met. In addition, smart meters record customer water usage, providing a clear picture of water consumption and conveying data to both consumer and utility, allowing for improved water management.

THE POWER OF DATA PROCESSING

Machine learning and artificial intelligence (AI) are used to process the sensing and data from other technologies but also to optimise the workforce and ensure customer needs are met. AI technology can recognise patterns in data and "learn" over time, updating algorithms as new information is presented. When paired with software as a service (SaaS) platforms, sensors, and communication networks, AI allows for the strategic and cost-effective operation of utilities, including better planning and execution of projects, better tracking and understanding of resource-loss in real time, more efficient collection and distribution networks, and maximum revenue capture and customer satisfaction. In this way, machine learning/AI technology help address the key issue of being data rich yet information poor in the water industry. Other AI services include chat bots which can be used to answer customer inquiries on demand, ensuring reliable, 24/7 service and enhancing customer satisfaction.

AUGMENTED, VIRTUAL, AND DIGITAL TWIN REALITY

Augmented and virtual reality (AR and VR) technologies provide their own, unique contributions to digital water. AR and VR technology has the potential to support decision-making in the field by providing holographic representation of pipes, cables and other assets, and offering immersive, scenario-based training for employees. Digital twin technologies merge GIS, sensors, and VR applications to generate working replicas of physical systems that combine physical data (satellite images) with real-time, in- situ data (sensors/internet of things) to simulate utility functions. Digital twins provide utilities with an ability to visualise and monitor current conditions as well as ask questions and predict real-world scenarios.

BLOCKCHAIN APPLICATIONS FOR WATER

Blockchain applications have the potential for direct, secure transactions between resource providers and consumers, peers, utilities and other players in the water sector. There are already several blockchain projects and trials occurring in the water industry, some of which are in conjunction with energy applications. Examples include: a project to integrate distributed energy and water systems in Fremantle, Australia and an Australian government funded project to develop a blockchain enabled system to monitor water trading and automatically update state registries. In the US, a water treatment technology provider announced it was creating a new blockchain protocol to make payments for international water treatment plant developments.

Taken together – remote sensing technologies, advanced in-situ sensors, AI, machine learning, AR/VR, digital twins, and blockchain – are the foundation of what digital water is, and as new, digital technologies are emerging, several market players, organisations and associations are emerging as key stakeholders, representing who digital water is.

The ecosystem of digital water

Water and wastewater utilities are at the centre of a greater ecosystem of digital water comprising their value chain and associated stakeholders. Each utility is unique in its digital journey and needs, and therefore has its own tailored ecosystem of digital water stakeholders. Generally, each ecosystem includes stakeholders from across the water and wastewater spectrum, such as private and public utility peers, governmental bodies, technology solution providers, academic institutions, consultancies, industry associations and technology accelerators. How a utility interacts with its ecosystem varies depending on its digital adoption phase, but it is evident that as utilities adopt digital solutions, they are evolving from having a limited, linear ecosystem to one that is more complex and interconnected, as outlined in Figure 1.2.

As shown in Figure 1.2, the 'Transactional' digital water ecosystem is the most basic model and is characterised by ad-hoc engagement with digital water technology solution providers, industry associations, academics, etc. As utilities begin to embrace digital water solutions, transitional ecosystems are formed by utility leadership (e.g., CEO, Board, etc.) or utility functions (e.g., IT, procurement, etc.). At the next stage of evolution, leading utilities have an established dynamic and fluid ecosystem of digital stakeholders where utilities become catalysts for mobilising these stakeholders to collaborate, pilot, and scale these impactful digital solutions.

No matter the complexity of water or wastewater utilities' ecosystems, there are a handful of stakeholders cited as central to a successful ecosystem. For example, global water infrastructure and technology companies hold a primary role within digital water ecosystems as they are the experts with market-tested solutions. Business solutions providers, SaaS providers, and communications providers comprise just a few examples of other companies a digital utility interacts with. Likewise, utilities are increasingly partnering with start-ups and water technology hubs and accelerators as a means to generate and benefit from innovative platforms and new technologies.

Academic institutions, industry associations, and water technology hubs and accelerators hold a stance as primary stakeholders due to their role in bringing together collaboration on and learning around novel research of methods and technologies as well as providing tools and platforms for adopting such technologies. Also, there is an ecosystem of investors, including angels, venture capital, family offices and socially responsible focused funds that are engaged in identifying and scaling innovative digital water technology solutions. Not to be forgotten is the role of the public sector in the water and wastewater utility's digital ecosystem. In most countries, local and federal governments provide the ultimate oversight over utilities, generating the standards and regulations that dictate how a utility conducts its operations. Water and wastewater utilities are constantly interacting with the public sector regarding funding and industry requirements.

Navigating the definition of digital water

This broadened view of 'digital water' – covering the organisational structure of a digital water and wastewater utility, the landscape of upstream and downstream end users, and the expansive digital water ecosystem – can at times seem daunting to navigate. However, utilities are not alone. Through engagement in professional and industry associations, and in peer-to-peer dialogue with other water and wastewater utilities, this broadened view turns from a web of complexity to a map of opportunities. In successfully navigating this map, there is the potential to transform the economics of water and wastewater management.

Transactional

Weak or ad-hoc links with organisations that focus on digital solutions driven by scoped projects or needs from within the water utility.

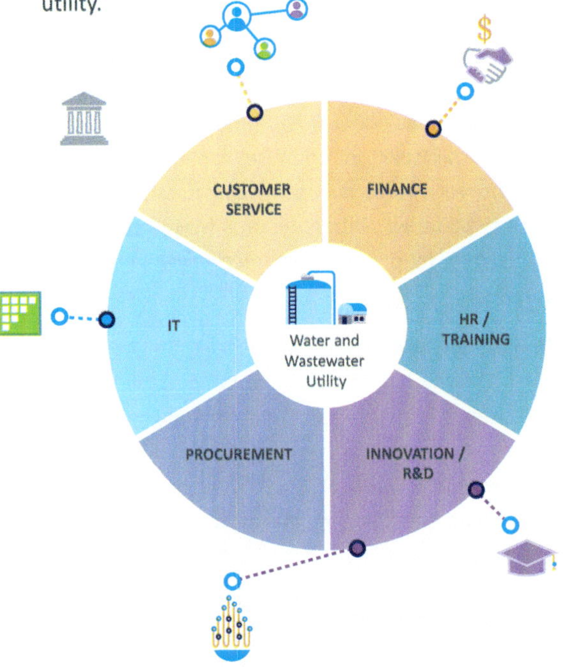

Transitional

Stronger linear relationships with digital solution providers that are driven by utility leadership or key utility functions such as procurement and IT.

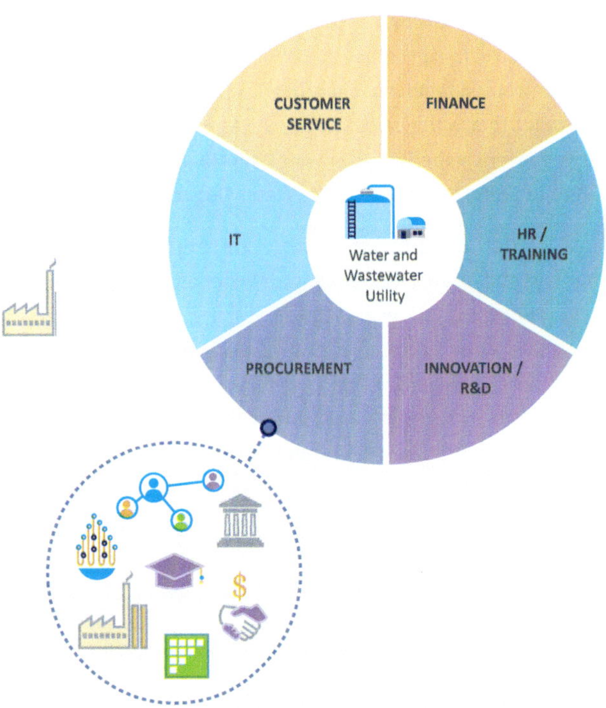

Dynamic and Fluid

Open engagement ecosystem with digital inputs from across stakeholders both external to the utility and with other utilities.

Peer Group & Network

Investors

Industry Associations

Academic Institutions

Water Tech Hub

Technology & Business Solutions

Public Sector

Note – the lines are representative and not a complete map of the relationship between stakeholders

Figure 1.2 Evolution of the Digital Water Ecosystem

1.2 Digital impact on the economics of water and wastewater

Taiwan has a history of drought and water shortages, an especially severe drought in 2002 triggered the Taipei Water Department to turn to digital solutions. "Since [the turn to digital solutions such as] the use of sensors, smart meters, and pressure control systems has improved water conservation, providing relief to a water-stressed city. As a result, the greater Taipei area has not experienced a water shortage in 17 years," states Chen Jiin-Shyang, CEO of Taipei Water Department.

The Taipei Water Department is just one example of the numerous compelling business reasons for water and wastewater utilities to adopt digital technologies. Scarcity, security, and resilience have now become critical drivers for water and wastewater utilities to deliver services to the public and private sectors. With this in mind, it is important to focus the limited capital and labour on the quality and reliability of essential services, while simultaneously investing in technologies, human resources, and initiatives to meet future demands. This makes it crucial for utilities to ensure that every dollar is being spent in the most effective way possible – maximising the value each project delivers to the utility, its customers and society.

The value case for digital water

The value created by the use of digital technologies across this expanded view of the water and wastewater utility, as captured in the previous section, is diverse. Often cited sources of value include 'decreased operational expenditure,' 'increased workforce efficiencies,' 'increased customer engagement and satisfaction,' and 'becoming an industry leader'.

As illustrated in Table 1.1, the potential sources of value created by the adoption of digital solutions are diverse and impactful, across the utility, the community and the environment. Within the utility, the value of digital solutions is not only felt across operations and financials, but also across areas such as the workforce and the utility brand.

Of course, each utility situation is unique, and each digital solution has a unique set of value drivers for the utility in both the near- term and the long-term. Below are a few examples of digital solutions which deliver impact across these areas of value.

COMMUNITY BENEFITS

- **Increased affordability:** Optimised capital and operating expenditure (as addressed by the digital solutions listed above) – combined with digital solutions such as customer-centric data analytics and scenario modelling for affordability – holds the potential to build a rate structure with long-term affordability, reduced bill shock and non-payment, and reduction in cut-offs to vulnerable customers.

- **Customer experience:** Similar to 'reduced operational expenditure' and 'increased capital efficiency,' nearly all digital solutions – which improve utility operations and financials – have a positive impact on the customer experience. Example digital solutions include a real-time digital twin of the water distribution network to optimise capital expenditure and reduce the volume of disruptive construction projects, and Advanced Metering Infrastructure (AMI) combined with data analytics to provide customer engagement on water consumption.

- **Environmental protection:** Efficiencies in operations (as addressed by the digital solutions listed above) – combined with technologies such as a real-time digital twin of the wastewater collection system, intelligent pumping systems, and a network of sensors in the watershed can ensure minimal contamination and maximised conservation of our watersources.

OPERATIONAL BENEFITS

- **Process excellence:** Digital solutions leveraging sensors, intelligent equipment, edge computing, and AI optimise individual components of the water and wastewater utility operations as well as connected processes across the water and wastewater utility value chain.

- **Predictive maintenance:** Digital solutions such as algorithmic and in- situ leak detection, asset management platforms, and AR / VR provide preventative and predictive maintenance capabilities that reduce downtime of critical assets and maximise effectiveness and efficiency of operations.

- **Regulatory compliance:** Digital solutions such as a real-time source-to-tap digital twin, powered by online monitoring and water quality models, and decision support and scenario modelling helps drive increased regulatory compliance across the water utility value chain.

FINANCIAL BENEFITS

- **Reduced operational expenditure:** Nearly all digital solutions have an impact on reducing operational expenditure. Example digital solutions include intelligent equipment that self-optimises for lowest energy use, real-time digital twins of the treatment plant to optimise energy and chemical use, and data analytics and decision intelligence platforms that enable efficiencies in decision-making.

Community Benefits

INCREASED AFFORDABILITY
- Improved long-term affordability of rate structure
- Greater transparency in the use of proceeds from water tariffs
- Reduced likelihood of bill shock, non-payment and cut-offs

CUSTOMER EXPERIENCE
- Increased customer engagement and responsiveness to customer inquiries
- Reduced disruptions in water service
- Reduction in the volume of disruptive construction projects

ENVIRONMENTAL PROTECTION
- Reduced risk of sewage overflows into the environment
- Reduced GHG emissions from utility operations
- Improved conservation and management of critical water resources

Operational Benefits

PROCESS EXCELLENCE
- Data-driven operations and decision-making reduces errors
- Speed in decision-making due to efficient data analysis and processing

PREDICTIVE MAINTENANCE
- Reduced number of emergency call-outs
- Reduced downtime of critical assets

REGULATORY COMPLIANCE
- Reduced incidences of failure and overflows
- Reduced risk of non-compliance resulting from network water quality issues

Financial Benefits

REDUCED OPERATIONAL EXPENDITURE
- Optimised operations reduce energy and maintenance costs
- Reduction in costs and risks associated with ad-hoc field maintenance

INCREASED CAPITAL EFFICIENCY
- Improved cash flow as a result of targeted rehabilitation of faulty infrastructure
- Reduced liability and costs from unexpected water main breaks and sewage overflows

INCREASED REVENUE
- Targeted interventions with faulty meters increases revenue
- Value-added digital services available to bulk water customers

Long-term Resiliency Benefits

INCREASED RESILIENCE
- Improved operational flexibility from changing climate and demographics
- Increased safety through rapid customer engagement on public safety concerns

WORKFORCE DEVELOPMENT
- Improved cross-department collaboration through systems integration
- Reduced safety risk to workforce through fewer emergency call-outs

BRAND AND INNOVATION
- Elevates utility brand and engagement in the water industry
- Enables the utility to more easily pilot and adopt latest technologies

Table 1.1 Digital Water Value Creation Overview

- **Increased capital efficiency:** Similar to 'Reduced operational expenditure,' nearly all digital solutions have an impact on increasing capital efficiency. Example digital solutions include algorithmic and in situ leak detection technologies that result in targeted pipe replacement and real-time digital twin of the wastewater collection network to optimise existing assets and avoid capital-intensive construction projects.

- **Increased revenue:** Digital solutions such as AMI and advanced data analytics of the metering network. These solutions can increase meter accuracy, maximise billing potential, meet and exceed customer's needs, and provide the opportunity to sell value-added services.

RESILIENCY BENEFITS

- **Increased resilience:** Digital solutions, such as a dense network of sensors, intelligent equipment, real-time source-to-tap digital twin, and data analytics and advanced simulation tools, enable a utility to be better prepared to their changing environment. Furthermore, the incorporation of external data sets such as weather and traffic data can improve a utility's ability to adapt operations to changing climate and demographics.

- **Workforce development:** Similar to 'customer experience,' nearly all digital solutions – which improve utility operations and financials – have a positive impact on the utility workforce. Example digital solutions include systems integration across data siloes, which improves cross-department collaboration, data analytics and decision intelligence tools that enable operator piece of mind, and predictive maintenance solutions (as addressed by the digital solutions listed above). The latter, in turn, decreases the need for emergency call-outs.

- **Brand and innovation:** Similarly, nearly all digital solutions have a positive impact on the utility brand and the potential to adopt latest innovations. As adoption of the variety of digital solutions increases, so do the capabilities and culture of utilities, ultimately enabling them to more quickly extract value from future innovations.

As many leading utilities around the world are already realising, the value created by the implementation of digital technologies is undeniable. According to Global Water Intelligence (GWI), potential savings on total expenditure over 5 years (2016—2020) globally for drinking water treatment, distribution, and customer services, metering and billing is about USD 176 billion, while the potential saving in the waste water sector is about USD 143 billion.

The transformative potential of digital water

Digital technologies have the potential to transform the economics of the water and wastewater sector. Through process optimisation; workforce transformation; enhancing customer engagement; aiding regulatory compliance; increasing sustainability, resiliency and watershed connectivity; and ensuring public health, transparency, and proper governance digital technologies are generating direct savings for utilities and creating both internal and external value across utilities' supply chains. As digital technologies create disruption across the water sector, the digitalisation of water and wastewater utilities will accelerate. The shifts driven by such digital transformations, however, will span wider than traditional utility operations, also impacting the nature of workforce operations, the role of utilities as a part of sustainable cities, the potential of green infrastructure, and customer-utility relationships. Furthermore, digital technologies improve day-to-day water management and build long- term resilience to disasters and climate change. Such improvements lead to increased water security for the industrial, commercial, agricultural, and domestic sectors, thereby having a direct impact on economic security and growth.

These deeper shifts include:

THE 'NO-COLLAR' WORKFORCE

The emerging 'no-collar' workforce means redesigning jobs and reimagining how work gets done in a hybrid human-and-machine environment. The development of digital technologies now requires the utility workforce to adapt and learn new skills in order to keep up with the pace of evolution within the global economy and systems of commerce. In addition to recruiting new talent proficient in information technology, companies need to train existing employees and attempt to continue to operate and adjust to new systems seamlessly.

Another way to frame the digital workforce is how the 'no-collar' workforce will be incorporated into company operations. In this scenario, robotics and AI will probably not displace most human workers. Instead these tools offer opportunities to automate some repetitive, low-level tasks. Perhaps more importantly, intelligent automation solutions may be able to augment human performance by automating certain parts of a task, thus freeing individuals to focus on more human-necessary aspects, ones that require empathic problem-solving abilities, social skills and emotional intelligence.

VR and AR applications can also benefit the water utility workforce by reducing risk and saving billions in maintenance costs, engineering tests and innovation, and allow users to test or simulate real-world situations without the usual dangers or costs associated with large engineering projects. With VR, asset maintenance professionals can immerse themselves to fully and accurately experience what a situation would be like in real life. VR also allows the identification of design flaws or other potential problems with efficiency, which can then be solved before any problems actually occur.

The demand for workers in the water utility industry is growing at a rapid pace due to the many applications of digital technologies. The role of businesses and of workers has changed and will continue to evolve with water resource management. Human capital is necessary in both physically developing water infrastructure networks as well as installing digital technologies. The ability for workers to interact with VR, AR and AI technologies is crucial and will create a multitude of possibilities for the sector to expand and become more sustainable for future generations.

RESILIENT AND SUSTAINABLE CITIES

Water utilities and cities are looking for ways to become more resilient to the impacts of increasingly frequent and severe floods and droughts. Losses due to disasters from natural and man-made hazards are mounting and on average cost governments over USD 300 billion globally each year. Some companies are rising to meet these challenges by providing services through monitoring flooding likelihoods and impacts in real time, helping to avert the human and economic costs of flooding, as well as assist in the aftermath.

Smart storm water systems are increasingly becoming available that leverage existing infrastructure. Furthermore, meeting conservation requirements in times of drought is becoming easier thanks to accurate groundwater resource modelling, for example, along with improved conservation habits.

Digital technologies will play a role in planning and redesigning cities to be more resilient. Remote sensing technologies for flood prediction and comprehensive design tools for hydraulic modelling are now available to manage storm water runoff and flooding from extreme weather events. Urban water systems – often vulnerable to extreme weather events, resulting in significant impacts to clean water distribution, wastewater treatment and storm water management – are also adapting microgrid strategies from the power sector. Water microgrids or 'micronets' provide redundancy, fortify against vulnerabilities, and can secure the resource supply chain.

DIGITAL AND GREEN INFRASTRUCTURE

Digital technology can provide real-time monitoring on green infrastructure performance. Cities and utilities that invest in green infrastructure may need a way to capture performance of investments at scale. For example, the City of Chicago, Illinois, wanted to better understand the performance of their bioswales (landscape elements, e.g., ditches, designed to concentrate or remove debris and pollution out of surface runoff water) and porous asphalt during storms. A team was assembled with the goal of better understanding the performance of the city's green infrastructure investments. Monitoring devices were installed on four green storm water management solutions: a bioswale; porous asphalt; a permeable paver and infiltration planter duo; and a permeable paver and tree-pit filter pairing. Over an 18-month period, cloud-based storm water management analysis and control software were used to gather live data on several green infrastructure storm water solutions. During this time, the monitoring system not only provided information to researchers and the city, but was also relaying data on the test areas' performance to the public.

Another example is from Ormond Beach, Florida, where real-time data analysis kept flooding at bay by automating water levels in storage ponds. Before Hurricane Irma hit Florida in 2017, Ormond Beach installed sensors on five lakes that are prone to flooding. When the sensors and software used detected an incoming downpour, the system automatically drained down the storage volume in preparation. The weather-respondent valves on the lakes' drainage system kicked in before the rain hit and helped prevent flooding during the storm.

THE SMART HOME AND CONSUMER

Digital solutions will also change the relationship water providers have with customers as society increasingly embraces digital technologies in all aspects of their lives (e.g., mobility, communication, and entertainment), and it is reasonable to conclude service providers such as water utilities will now be part of the mix.

With new efforts toward sustainability and water conservation efforts, water utility companies are beginning to establish innovative strategies to help engage consumers and restructure the way people think about water use. Research and case studies have shown that consumers are more likely to change their water usage when new strategies are easy to install and access and when water-saving efforts do not considerably change their daily living habits.

The concept of a 'smart home' opens up a wealth of new opportunities for water sustainability. The growing range of technological home applications combines increased awareness with convenience to reach sustainability. For example, a study by Singapore's national water agency PUB found that a person could save up to 5 litres of water a day using smart shower devices. In another example, the Mackay Regional Council in Queensland has introduced automatic water meter readings to empower customers to better manage their water consumption and save money. The digital meters also alert customers and local authorities to leaks. Having the technology can allow local authorities to repair the issues within a much shorter timeframe, reducing water wastage.

Companies that take advantage of these developments in customer service are benefiting. With new digital technologies such as AI chatbots, customers can ask questions and get answers whenever they want, opening vast possibilities for consumer engagement, providing customer alerts, and also water consumption and conservation information. Utilities that embrace these technologies are improving their customer service and meeting the high demands of consumers.

Beyond the utility

Digital technologies have exponential growth in capabilities and performance and exponential decline in costs. As a result of these attributes, digital technologies are rapidly scaling in developed and emerging economies.

The adoption of exponential technologies allows countries to leapfrog past last century solutions to off-grid, decentralised, and distributed water and wastewater infrastructure. Likewise, the trend of increasing adoption of smart devices and online transactions is accelerating and becoming something that is more and more expected by customers and other stakeholders.

These digital transformations within related industry sectors will not only impact the uptake of digital technology by water and wastewater utilities, but will further enable the digital disruption of the water sector.

Digital transformations are already occurring across the market in the ways customers interact with goods and service providers. Customers are increasingly demanding user-friendly, 24/7, multifunctional service; easy access to information; and immediate transaction abilities. Flexibility and on-demand capabilities are increasingly both expected and assumed. With the ability to manage their finances (e.g., online banking), shop (e.g., Amazon), order food (e.g., DoorDash), monitor their homes (e.g., Nest), and schedule transportation (e.g., Uber) already in the palm of their hands, it is only a matter of time before customers realise they have the power to demand the same level of convenience via digital technologies from other sectors. As more industries join the transition to digital, a foundation

and culture already embedded in digital technologies holds the potential to better prepare utilities for the future.

Individual industries and companies, however, are not the sole instigators in the shift to digitalisation. Emerging smart city initiatives are creating demand for digitalisation across industries. As cities aim to optimise infrastructure, industries and services through increased connectivity and better engagement of governments, citizens and businesses, digital solutions will need to be at the forefront of meeting new challenges. In that context, the very nature of a water utility can be used as a springboard to a 'Smart City.' Water and sanitation are among the most essential services a city provides and are at the foundation of economic stability. In developed countries, water utilities touch virtually every citizen, home and business, meaning cities can exploit a digital water utility's communication network, customer base, and immediate value propositions to demonstrate and communicate the overall benefits and successes of the city being an interconnected enterprise. As urban evolution is pushing cities toward becoming intelligent, connected, ecosystems of sensor-based infrastructure, utilities that have already embedded digital technologies into their operations will be better equipped to fulfil new requirements and hold an active role as a part of smart city initiatives.

New and emerging markets also provide incentives for water utilities to adopt digital technologies. As countries expand and improve upon water services in Asia and the Pacific, regions where the presence of microelectronic and pharmaceutical industries mean digital automation is already highly adopted, digital technologies are likely to be expected and demanded as the water sector expands. Likewise, as the world aims to meet Sustainable Development Goal (SDG) 6 by 2030, new markets in Asia, Africa, Latin America, and the Middle East will create a niche for technology to meet rising demand in dense, urban areas and water-scarce regions. In North America and Western Europe, where extensive infrastructure systems already exist, markets for digital technologies are emerging to address the issues associated with aging infrastructure. Overall, a study by GWI predicts global demand for control and monitoring solutions will rise to USD 30.1 billion by 2021, a market that utilities with pre-existing digital platforms will be better prepared to embrace.

Richard Appiah Otoo, Chief Technology Officer of Ghana Water, said it best when describing his utility's transition to digital water, "The world is moving in the direction of technology and Ghana Water cannot afford to be left behind." Ghana Water is one of several utilities worldwide that has recognised the value of adopting digital technologies to prepare for future markets and meet consumer demands. Utilities, like Ghana Water, who start their digital journey now will be better prepared to extract even further value from digital water in the future.

Digital technologies alone will not only create new value for water and wastewater utilities but they will be an enabling force for the adoption of other technologies such as new water source collection (e.g., air moisture capture, water reuse/recycling, etc.) and decentralised water treatment systems (e.g., building and community scale) and actively shift communities from exclusively centralised systems to hybrid solutions that incorporate the advantages of more traditional systems and innovative methodologies powered by digital technologies.

To further highlight the potential of incorporating digital technologies into the water sector, reference can be made to the energy sector's adoption of microgrids and smart-sensing to reduce vulnerability of large-scale plants to climate effects, decrease operational complexity, and provide resilience, reliability, flexibility and redundancy to the sector. These benefits can also be seen through establishing micronets (water microgrids – decentralised water and sanitation systems). By downscaling plants and treatment facilities to serve smaller regions, monitoring and maintenance becomes easier and less costly, and the installation of new digital technologies is more realistic.

The ultimate reason for heightened interest in scaling digital water technology solutions is the urgent need to ensure access to water for economic development, business growth, and social and ecosystem well-being in the face of increased demand for water and the impacts of climate change. The role of the water utility sector in addressing these needs has never been more critical.

1.3 Navigating the digital journey

As utilities implement new digital solutions and update their business models to embrace the digital era, they find themselves at varying levels of digital maturity. There are various types of digital adoption including adoption of analytics to create value out of existing data; adoption of hardware and software to create systems (simulations) based solutions; and building of communications and IT infrastructure, and human capital development to create a "pull" for smart systems.

The digital water adoption curve

Utilities are at different levels of maturity in adopting these categories of digital solutions and approaches. To understand the state of digital maturity in the water sector, we elicited insights from 40 leading utilities worldwide, speaking to 15 utility executives through in-depth interviews and the remainder through a detailed survey process. Figure 1.3 shows the global spread of utilities whose input informed our research and aided the development of this report. Please see the acknowledgements for specific utility contributions.

The Digital Water Adoption Curve, in Figure 1.4, adapted from Gartner 2017, is a synthesised view of how utilities are adopting digital technologies. Generated to be a working tool for utilities now and in the future, the Digital Water Adoption Curve provides a means for utilities to assess where they are in their digital maturation and to have a general roadmap on where to head next.

The curve begins with utilities at an immature digital development phase. It then expands through utilities that have become digitally aware or that have incorporated digital technologies within and between their processes and moves on to utilities with an agile, innovative business structure that have fully embraced digital technologies.

The maturation of a water or wastewater utility along the curve is shown as a utility progress from having little to no digital infrastructure to having opportunistic, systematic, and transformational digital systems and strategies. In the interviews and surveys of leading utilities, executives were asked to reflect on their organisation and assess their own phase of digital maturity. Responses spanned across the entire Digital Water Adoption Curve spectrum, with some utilities having more conservative beginnings and others already largely embracing the full expanse of digital technologies. With an average adoption level of 'opportunistic,' it appears that many of the utilities surveyed and interviewed have started their digital water transformation journey.

Those utilities in early development stages are focusing efforts on implementing software platforms (National Water and Sewerage Corporation), new sensors and smart meters (Shenzhen Water Group-China). Likewise, increasing automation for remote control (Berliner Wasserbetriebe-Germany), combining networks (Umgeni Water-South Africa) and enhancing internal infrastructure remain top priorities. Utilities further along in their digital maturation have already incorporated technologies like VR and big data into automated processes and decision-making, helping to run smart solutions (Macao Water-China). Others have expanded beyond their organisation to provide services and support to external utilities (AGS Water-Portugal).

Nevertheless, all utilities have room to grow and as future conditions (climate, population, demand, etc.) change and technologies continue to evolve, there will be unending opportunities for the adoption of new and improved digital infrastructure. Utility leaders interviewed shared their efforts

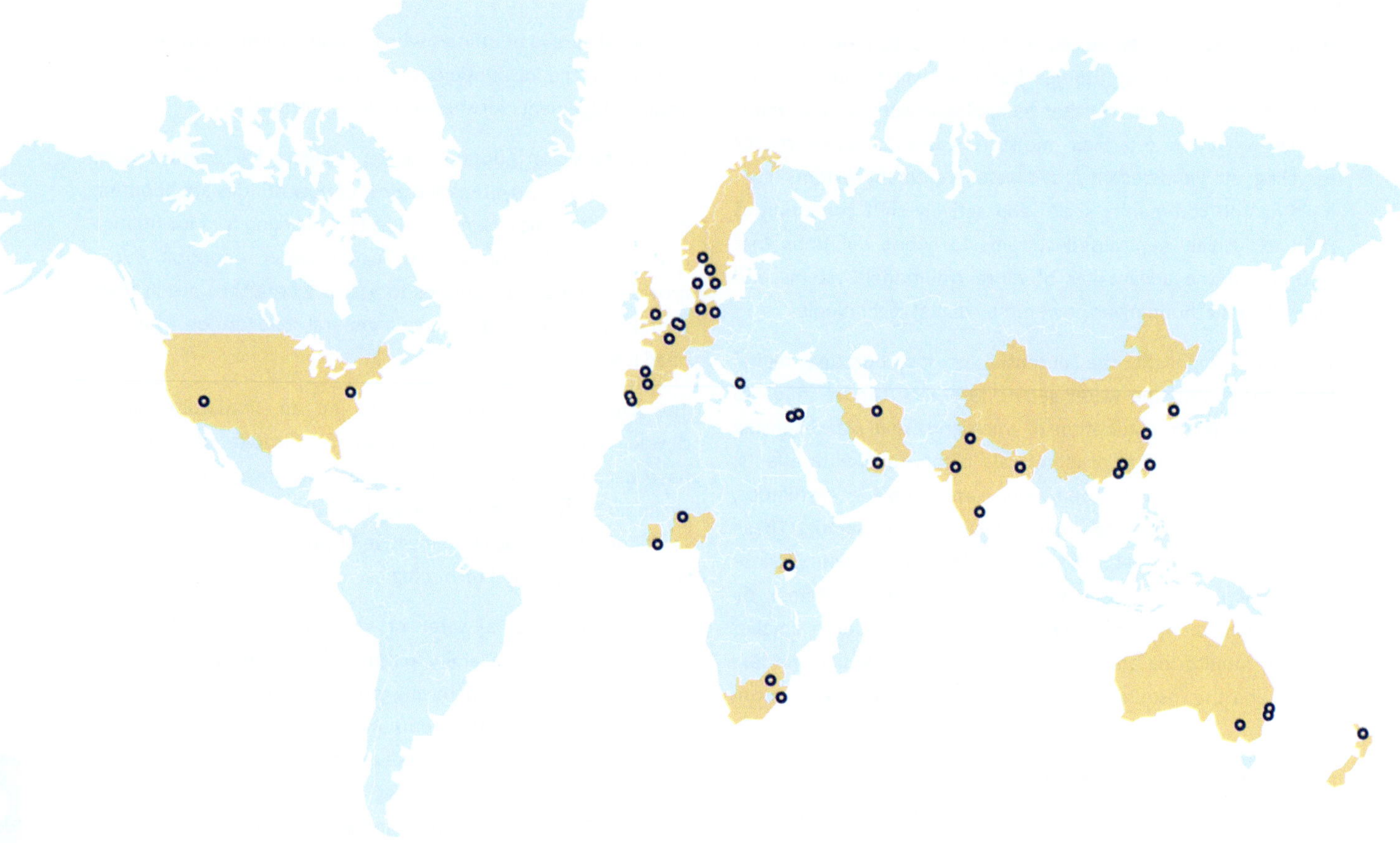

Figure 1.3 Geographical span of utilities interviewed and surveyed

and advice on how best to further advance along the Digital Water Adoption Curve, revealing necessary steps for embedding digital technologies within utility operations, providing insights on the source of ambition for digitalisation, and sharing critical lessons learned that helped to initiate and further propel them along their digital journey.

LESSONS FROM UTILITY PEERS

The digital water transformation is here. Nonetheless, moving away from traditional methods and infrastructure requires effort and commitment. Based on the input from those interviewed and surveyed, six overarching utility actions have been identified to accelerate the utility journey across the Digital Water Adoption Curve:

Set the ambition at the CEO and Board level: Having the support and leadership of the utility's executive team and board is a critical accelerator to the implementation of digital technologies. Executive barriers will be some of the hardest to overcome, yet, since the digital transformation of utilities will require organisation-wide changes in operations and strategy, board authorisation is key for a utility to embark on their digital water journey. At the Las Vegas Valley Water District, where

board leadership has already adopted company-wide goals for innovation and digital solutions, David Johnson explains that, "As a public utility, getting our board to adopt [digital] goals and make them a priority opened up the pathway for us to be able to allocate funding toward projects." He continued by sharing the observation that barriers throughout their digital water journey have been significantly reduced due to strong board level leadership. Consensus was reached across utility executives that top management must understand the risks and benefits of digital technologies, as well as both support and take the lead in driving the adoption of big data and digital infrastructure for projects to be successful.

Build a holistic digital roadmap: As the utility embraces change and begins its digital water journey, it is helpful to have both a strong roadmap for digitalisation and a clear business strategy. Communication and awareness of the direction the company is going are essential. Consensus must be built within the utility on how the digital journey will unfold, and maintaining the customer and business outcomes as focal points throughout the digitalisation process is fundamental. Silver Mugisha, Managing Director of National Water and Sewerage Corporation explained that their "biggest success has been on how we have managed to integrate technological innovation throughout our business

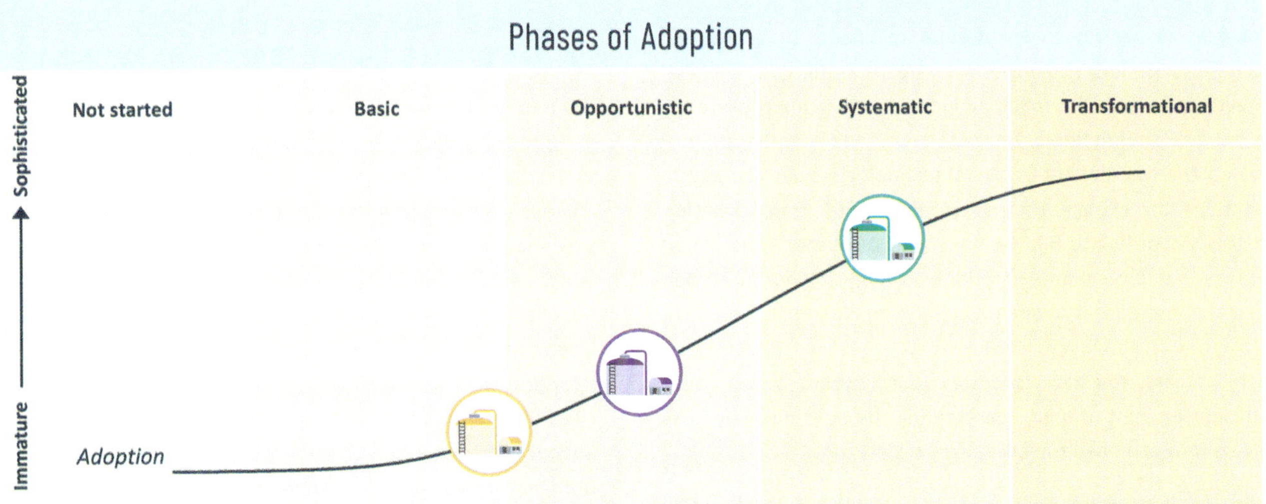

Example Characteristics

• Traditional, legacy analogue infrastructure • No digital strategies or technologies	• Begin incorporating digital technologies into operations • Develop online monitoring capabilities, i.e., internet of things, SCADA	• Most operations have been redesigned with digital automation and control • Analytics tools utilised for process optimisation	• Digital technologies are well established • Inter-process automation/control • Internal resources and platforms developed for working with digital infrastructure	• Digital technologies incorporated across business and operations processes • Advanced analytics used for decision-making

Actions to Move up the Adoption Curve

• Acknowledge digitisation as a business priority • Develop a digital strategy within top management • Pursue pilot projects to explore digital implementation	• Mobilise pilot projects and learn from industry peers and research • Transition recording, billing, etc. from paper to digital • Ensure customers and employees are well-informed on utility direction	• Solidify data infrastructure to enable next-level digital tech incorporation • Align utility around data-driven goals	• Develop innovative, new products and services through digital technology • Provide resources to industry peers • Have an evolving digital framework • Ensure projects align with digital goals and business strategy	• Continue pursuing increased efficiencies via digital technology • Continue learning from industry peers, conferences, etc. • Exchange best practices with other utilities.

 **UTILITY EXAMPLE 1:**

- Has deployed online sensors across some parts of its network to monitor water quality
- Is developing a predictive catchment model for supply areas
- Commenced pilot testing stage for the implementation of other digital projects

 **UTILITY EXAMPLE 2:**

- Has implemented SCADA systems, automated processes, and deployed technologies for remote monitoring
- Use GIS, hydraulic networking models, and Virtual Intelligence to inform decision-making
- Using cloud services, digitised billing, merged databases, developed a customer service app, and enhanced inter-office communication

 **UTILITY EXAMPLE 3:**

- Have become an internally digitised company with well established digital technologies
- Collect actionable data using digital technologies and use virtual intelligence and data analytics tools to make decisions
- Have developed services and tools to support both their group and external utilities
- Use digital technologies to manage work orders, conduct real-time monitoring, detect and respond to events, optimise processes, generate reports, etc.

Figure 1.4 The Digital Water Adoption Curve

processes, especially in an attempt to create an everlasting customer experience". Further, to prepare a utility for success, it needs to be sure the roadmap includes educating consumers, politicians, shareholders, management, and employees not only on the cost-benefit of digital technologies, but also on its intentions for change throughout the digital transformation process. According to Biju George, from DC Water, "The digital strategy has to become a corporate strategy. It's not an option to sit there and let it happen, you have to plan for it. You have to train your employees towards that, you have to relook at every process. You have to design your systems to give you a sufficient amount of data, representing the right diversity (less correlation) you need to make efficient decisions."

Build an innovation culture: New technologies and digital solutions are being innovated across the water and wastewater utility value chain. To begin identifying, evaluating, and exploring these technologies, there needs to be an organisational curiosity for new technologies. Operators, IT staff, Finance, Technicians, Executives, and others have to be the scouts for initial sourcing.

Digital solutions are continuously evolving and so too must water and wastewater utilities. Exploring and adopting these digital solutions enables a culture of innovation, and once the latest digital solutions are mastered, the utility will be ready to begin the next level of business transformation, thereby creating a cycle of continued digital maturation. Claire Falzone, CEO of Nova Veolia-France, emphasised a need to become more customer-centric as well as being prepared to continuously adapt and evolve – using new technologies – to current and future water challenges.

Leverage pilots for an agile mindset: Pilot projects offer a means to explore new technologies and have a more holistic understanding of their physical and financial effects on operations before committing to large-scale implementation. As Dr Hamanth Kasan, General Manager of Scientific Services at Rand Water-South Africa notes, case studies, pilot projects and a track record of successes shown by initial testing during pilot projects will help build momentum for moving utilities up the Digital Water Adoption Curve.

Develop architecture for optimising data use: The data collected through digital solutions is only useful if you can structure and extract value from it. Developing a data warehouse, where operational data sets become available to finance, engineering and IT specialists who can use the data to optimise business processes, is a key step in effectively digitalising utility infrastructure. Meriem Riadi, Chief Digital Officer of SUEZ Group, says, "Advancement requires aligning [the] utility around data – converting data to intelligence and business use cases – and, more importantly, developing a culture of data (e.g., an understanding of data, its value and the multiple ways it can be used)."

Collaborate with utility peers: It is important to realise that, as you encounter challenges and explore new digital solutions, you are not alone. Whether you are a large or small utility, established or just emerging, or are located in a developed versus developing country, someone else has faced the same challenges within their digital journey as you. Fortunately, there is openness and a willingness to share information within the water sector and utilities should actively seek out these insights. After all, global utilities share the same goal: to provide safe, reliable water and wastewater services to everyone.

Dave Johnson of the Las Vegas Valley Water District cited the support from and their work with water tech hubs/accelerators and their ecosystem of stakeholders coupled with the ability to learn from the power sector (Nevada Power) as important in their digital water technology transformation.

In summary, the new era of digital water utilities is here and already evolving. Developing strong business cases and road maps for the digital journey will aid in gaining the trust and support of customers, shareholders, utility staff and politicians. Be open to learning from other utilities and the surrounding ecosystem, embrace innovation, and share information when possible. As Claire Falzone of Nova Veolia-France encouraged: "If you have any doubt, just try it. Try small at first. This is just the beginning of the digital water journey and if you don't adopt digital technologies, someone else will."

BUILDING DIGITAL INTO YOUR ORGANISATIONAL CULTURE

Digital technologies cannot be sought as simply surface-level solutions. To operate effectively, they must be incorporated into the very backbone of water and wastewater utilities. From physical infrastructure and business services to data management and customer relations, digital technologies can and should become interwoven with all levels of a utility's operations. The journey must begin somewhere, however, and our conversations with utility executives have identified three primary mechanisms by which digital tech can be built into 'organisational DNA'.

First, engaged leadership from within a utility's executive team is fundamental both to developing a foundation of digital technology and advancing an organisation along the Digital Water Adoption Curve. Executives as well as the oversight board must discuss the direction they envision for the utility and how digital technologies could fit within and enhance that vision. Identifying priorities, outlining strategies, developing roadmaps, and allocating funding specifically to digitalisation are critical steps any executive team must take for their utility to transition into the digital era. In tandem, board level approval of such goals, visions and budgets is necessary. Executives and boards alike must then hold each other accountable, ensuring goals are met, resources are allocated effectively, and the utility's mission is upheld.

Second, developing or expanding upon existing roles such as that of Chief Digital Officer (CDO) can ensure digital technologies remain a priority within top management and aid in enabling and accelerating the digital adoption process. The development of a CDO position at the SUEZ Group allowed a leading individual, Meriem Riadi, to create a digital team, develop roadmaps, study the trajectory of digital technology in the water sector, accelerate the delivery of digital projects, and work on innovation more efficiently with partners, adding overall value to the company and increasing the success of digital projects.

Digital projects, however, can still be instigated and gain traction regardless of an individual's role on the executive team or of leadership within top management. In Biju George's case, a single, motivated individual with a curiosity toward and drive for innovation can be just as influential from an operational or middle management standpoint. Before starting his position at DC Water, Biju worked in several roles, from engineer to management, in Cincinnati's Sewer and Water departments. A fascination with digital technologies led Biju to explore innovative, emerging solutions, ultimately leading Cincinnati Metropolitan Sewer Department and Water Works to become an early adopter of intelligence communications services: software programs that provide actionable information to the utility.

Biju took the initiative to expose employees to digital innovation, transforming their views on the adoption of new technologies. Likewise, he initiated a sewer digitalisation project and collaborated with vendors to develop products to fit his utility's needs. Meanwhile, Biju created a Watershed Operations division at the Metropolitan Sewer Department and recruited the organisation's Chief Information Officer (CIO) to run it – recognising that a technology expert would be best able to learn, engineer and deploy emerging digital technologies. As an executive at DC Water, Biju continued exploring and instigating the adoption of digital projects. With a track record of success long before his executive influence, however, Biju provides an example of how stamina and a vision for innovation from any individual willing to take on the challenge can propel a utility along its digital journey and embed digital technologies from within.

Third, digital projects can be woven throughout utility infrastructure from within in a bottom-up approach. At Umgeni Water, Dan Naidoo shares that the technical team was driven to pursue digital technologies largely by a need for operational efficiency, optimisation and increased resilience. Over time, as the use of digital systems and tools spread throughout the utility and returns on investment (ROI) were realised, the ROIs of previous digital projects became the catalyst for further digital investments. In addition, the expansion of digital infrastructure led to the merging of networks across departments as data was collected and shared within the utility. Mr Naidoo noted that, although projects initially were operationally, efficiency and financially driven, they have since grown within the organisation to reach the CEO/board level and digital transformation has now been embedded into the utility's business strategy at an executive level.

It is important that utilities use caution and thoroughly explore digital technologies (e.g., pilot projects) to understand their uses and effects on operations before implementing at a large scale. Water and wastewater utilities have a responsibility to their customers and digital technologies can have a direct impact on public health and economic stability in their communities.

Nonetheless, stakeholders interviewed indicated that the transition to digital technologies in the water sector is both inevitable and necessary. As future utilities begin their digital journey, in-depth interviews with utility executives show that digital projects can be instigated and embedded into a utility's backbone from any level within the organisation. Only when digital technologies are an integral part of water and wastewater utility's DNA, will we be able to solve water and meet the rising challenges and increasing demand of the next century.

1.4 Accelerating digital water adoption

Every day, billions of individuals go without access to affordable water and wastewater services, millions of litres of clean water leak into the ground, and thousands of litres of raw sewage are released into the environment. These statistics – combined with the significant value potential from adopting digital solutions – means that every minute counts in accelerating the adoption of these technologies. While the adoption of digital solutions will seldom be a smooth journey, there are key enablers that must be leveraged to the fullest extent by the water sector.

Key barriers for digital technology adoption

The digital water journey is one containing many hurdles and barriers that at best, slow the implementation of a digital solution, and at worst, prohibits even the piloting of a potentially transformative solution. To speed progress and fully realise the opportunity, there are regulatory, technological, and organisational challenges that must be addressed by the water sector.

SYSTEMS INTEGRATION AND INTEROPERABILITY

As seen in in previous sections, water and wastewater utilities are complex organisations with numerous data siloes. In addition, most utilities have legacy systems containing operations critical information, as well as valuable historical context on the changing urban watershed. Across these data siloes and electro-mechanical rotating equipment – often from various suppliers with various communication protocols – there is a growing challenge of systems integration and interoperability. There are solutions to bridge this systems integration, but open architecture and standardisation holds the potential to accelerate adoption of digital solutions.

HUMAN RESOURCES IMPACT

The success of digital solutions is often not a function of the technology, but rather of the people and the processes that leverage this solution. Adopting digital technologies can bring up human resources concerns related to skill gaps, workforce transition and change management. Digital solutions can also create value for the workforce by increasing workforce development opportunities and cross-department collaboration. For example, Hamanth Kasan (Rand Water) stated that "culture is important" and that utilities must "overcome fear of data and transparency". As young engineers enter the workforce, the willingness to explore digital technologies is growing. Another way to change culture and overcome the fear of transparency can come from CEO and board level commitment and a clear strategy.

A similar challenge and opportunity are utility operational silos. Richard Appiah Otoo (Ghana Water) said that "silo mentality was a problem" that had to be overcome in order to adopt digital solutions. Technology solution providers need to frame all digital solutions with the workforce in mind, and the broader water sector could benefit from additional research on the best practices in workforce development during the digital water journey.

FINANCING SOLUTIONS WITHOUT A CLEAR VALUE PROPOSITION

Digital solutions can deliver impact across a diverse set of value drivers, some of which are well-defined (e.g., reduced operational expenditure) and others that are less well-defined (e.g., increased resilience). With limited budgets, it can often be a tough decision between using budget towards a typical maintenance activity and deploying a digital solution that can drive long-term efficiencies in total asset management. Technology solution providers need to provide a clear definition of the total value created by digital solutions, and the water industry needs to provide additional case studies and proof points for how to account for the less well-defined sources of value. Additionally, new business models that better align the timing of value creation with capital expenditure on a digital solution are needed.

There is always a danger of wanting to do everything at once – 'build Rome in one day' explained Silver Mugisha of National Water and Sewerage Corporation. They had to agree on the priority systems and process that would have the greatest impact on their business by digitising them.

CYBERSECURITY

Cybersecurity and customer data protection are critical considerations when deploying digital solutions. To date, this barrier is being addressed by new technology solutions (e.g., cybersecurity systems) and by anonymising customer data to maintain privacy. However, continual advancements in technologies, standards and processes are needed to maintain security with our critical water resources.

While these barriers may at times seem insurmountable, talk to any utility executive and you'll find numerous examples where these challenges have been overcome. Take the Berliner

Wasserbetriebe as an example. Despite concerns on data, workforce and legacy systems, sewer maintenance at Berliner Wasserbetriebe is already semi-automatic and targeted towards digitally identified condition failures. Robots deployed in Berlin's sewer systems photograph infrastructure, which is then digitally recorded, analysed and sent to operators for maintenance approval – improving the speed and efficiency of addressing maintenance needs. This is just one example of many where perseverance, creative problem solving and executive leadership have been leveraged to overcome barriers in deploying digital solutions.

Accelerating forces for digital adoption

As seen in the previous sections, many utilities are still finding ways to extract value from digital solutions. This is the result of key enablers, which each utility is leveraging based on their situation and location, to override these barriers.

WATER REGULATIONS AND PUBLIC POLICY TO ENCOURAGE DIGITAL ADOPTION

New water regulations and public policies are emerging around the world in response to the new normal of prolonged drought (e.g., climate change) and resultant water shortages. For example, the California Sustainable Groundwater Management Act, which mandates the development of long-term water use strategies, is driving the development and adoption of cost-effective digital technologies to measure real-time water use. Local farmers, cities and water utilities alike must reduce water usage in compliance with new requirements under these laws. Such legal changes will force utilities in water scarce regions to turn to innovative technologies and business models to conserve water while continuing to meet demand. As another example, in the UK, the Water Services Regulation Authority (OFWAT) now mandates water companies to have at least five ways for customers to contact their utility, three of which must be digital. A similar OFWAT programme rewards water companies whose customers report high levels of satisfaction and penalise those considered to be under-performing, thereby incentivising process automation and the use of digital technologies to improve water and wastewater services.

DATA STRUCTURING SOLUTIONS FOR LEGACY SYSTEMS

Water utilities are now dealing with large volumes of data that are both structured and unstructured coming from disparate sources. Most utilities report that accessing data from legacy systems still presents a challenge. The key to maximising the use of big data is accessing the right data when it is needed by the applications. We see an increase in the use of application programming interfaces (APIs), which provide a way for retrieving data programmatically by any software application. Various software applications across the utilities can then use APIs to access the needed data from existing legacy systems, sensors and other applications regardless of data location, utility department or functionality needed. The same data sets can be used and reused for multiple purposes, thereby increasing the value of digital solutions.

DEMOGRAPHICAL SHIFT TOWARDS DIGITAL

Perhaps the most powerful enabler will be the digital customer and workforce. Generational changes will force the adoption of digital technologies – because customers and utility professionals will expect and demand that core services such as power and water embed digital innovations into their products and services. Couple this generational force with the emergence of the no-collar workforce, and the move to a digital water utility appears increasingly inevitable.

SITUATIONAL TRIGGERS FOR THE DIGITAL JOURNEY

Changes in a utility's situation – often triggered by an external event, such as a demographic shift, a significant flood or increased water scarcity – have been commonly cited as a catalyst for triggering the adoption of a digital solution. For example, growing demand in water users combined with labour shortages drove Shenzhen Water Group's digital agenda whereas expansion from metropolitan areas to more remote regions in South Africa fuelled Umgeni Water's desire for process optimisation. For many, the need to improve customer engagement triggered the start of a digital journey, while others were driven by industry competition and the fear of being left behind in a digital era.

Across the water and wastewater utility sector, there is a growing message of urgency to address water challenges and ensure adequate access to water and wastewater services worldwide. There's an imperative to maintain flexibility and develop an ability to adapt to growing populations, urbanisation and climate change. This in itself is becoming the catalyst for change and, moving forward, utility leaders, regulators, associations, etc. will be called on to take the necessary actions to guarantee reliable, sustainable water and wastewater services for their respective populations.

Based on our research, experience, and input from those interviewed and surveyed, eight overarching actions have been identified to accelerate the utility journey across the Digital Water Adoption Curve:

1. Set the ambition at the CEO and Board level: Utility leaders agreed that having the support and leadership of the utility's executive team and board is a critical accelerator to the implementation of digital technologies.

2. Build a holistic digital roadmap and a clear business strategy: Utilities must create internal consensus on how the digital journey will unfold, maintain the customer and business outcomes as focal points throughout the digitalisation process, and educate key stakeholders (consumers, politicians, shareholders, management and employees).

3. Build an innovation culture: Utility operators, IT staff, finance, technicians, executives, and others have to be the scouts for identifying new technologies. However, to drive adoption, utilities must focus on fostering an organisation-wide curiosity and competency for embracing digital innovation.

4. Leverage pilots for an agile mindset: Pilot projects offer a means to explore new technologies, build momentum, and create a more holistic understanding of their physical and financial effects on operations before committing to large-scale implementation.

5. Develop architecture for optimising data use: Developing a data warehouse, where operational data sets become available to functions such as finance, engineering and IT specialists who can use the data to optimise business processes, is critical to creating value from data and effectively digitalising utility infrastructure and connectivity.

6. Cultivate your digital ecosystem: Utilities should leverage insights on digital migration from peers, industry associations, academics and technology hubs/accelerators, who are further ahead of them on the Digital Water Adoption Curve. Fortunately, there is openness and a willingness to share information within the water sector and utilities should actively seek out these insights.

7. Embrace the digital water value case: The digital water value drivers within the utility, surrounding community, and in the long term, are diverse and transformational, resulting in a compelling case for accelerated adoption. The community, operational, financial, and resiliency benefits created by digital technologies generate exponential value for utilities.

8. The water sector needs to unite around solving key barriers: Key barriers such as interoperability, regulations, culture, and cybersecurity must be addressed by the industry as a whole.

Any platform for the adoption of digital technologies by utilities must begin with a thorough understanding of those technologies, the recognition of specific challenges faced by the utility, and a commitment to executing a strategy to address those challenges with new and innovative technologies and practices. It is important to remember, however, that the technologies will change and that the technologies themselves are not the solution. Rather, their implementation and the various ways in which they create value for a utility will be the solution to some of water and wastewater utilities' greatest challenges (e.g., non-revenue water, storm water and sewage overflow, etc.).

1.5 The roadmap forward

There is no question that the digital age has arrived. Digital technologies are now embedded in our daily lives transforming sectors such as communications, transportation, entertainment, education, manufacturing and healthcare.

The transformation is inevitable as water and wastewater utilities are now facing new risks from increasing demand, water scarcity, water quality and water security, exacerbated by aging and underfunded infrastructure, out of date public policies and climate change. The adoption of digital technologies will become increasingly necessary to provide improved, more reliable, secure, efficient, and cost-effective water and wastewater services.

While these risks can appear daunting for utilities, digital water technologies hold the promise of enabling water and wastewater utilities to make a much more profound contribution to sustained economic development, business growth and social well-being. It will now be feasible to create water abundance by deploying exponential technologies, of which digital solutions are key. This in turn will ensure SDG 6 can be achieved – securing water and wastewater services for all – and it will advance all other SDGs, which are water dependent.

Yet this digital transformation is not self-fulfilling. Digital water technology adoption requires the engagement and commitment of utility staff and customers as well as incumbents, start-ups and entrants from other sectors across the value chain. These diverse stakeholder groups are now converging in the water sector to scale digital solutions and catalyse the adoption of water solutions.

Digital technologies also bring new challenges such as cybersecurity. As a result, for innovative digital technology entrepreneurs to succeed, they must focus on integrating security into solutions, with systematic management of risks to mitigate operational network disruption risks and softer business network risks (theft or loss of data and damage to internal business systems).

The water sector faces a stark choice: resist the rise of digital solutions and struggle to adapt to water challenges, or fully embrace the digital revolution in collaboration with innovators to unlock a new era of water abundance.

Digital technologies are considered exponential. Exponential technologies (e.g., additive manufacturing, alternative energy systems and biotechnology) spur dramatic growth in capabilities and declining costs. However, the adoption of exponential technologies is challenged by linear thinking and experience, as well as how to prioritise and direct funding that can provide long-term solutions. As a result, what we observe with the digital transformation of water is a wide range of levels of adoption from early stage to advanced.

As water and wastewater utilities continue to mature into the digital era, there is a need for a smoother transition to digital technologies to ensure adequate services throughout the utility's digital journey. To aid in this transition, we tapped into the knowledge and experience of utility experts and executives, gathering novel research by conducting interviews and surveys with top water and wastewater utilities around the world. Through their insights and lessons, we were able to develop a roadmap for water and wastewater utilities as they begin and advance along their digital journey, as measured and guided by the Digital Water Adoption Curve. The Digital Water Adoption Curve is meant to be used as a tool for utilities both now and in the future. Other insights from this report provide additional support on how to begin and progress along this curve.

As the water sector embraces digitalisation, utilities must ensure the outcomes of digital projects remain focused on benefits to their business and services provided to their customers. This can be achieved by developing a digital strategy and roadmap, embedding it within the utility's business strategy, and ensuring it is well communicated and adhered to.

Gaining the support of top management (e.g., CEO, board, etc.) early on is critical in the development of digital projects as is developing or strengthening an IT/digital technology team.

Likewise, utilities must build a strong foundation in fundamental technologies with adequate data centres/ platforms for working with the bulk data digital technologies produce.

From there, utilities will be better prepared to expand upon digital infrastructure as new projects develop, new challenges arise, and new technologies emerge. Following these and other guidelines described earlier in the report is highly recommended as water and wastewater utilities join their industry peers in the digital era.

1.6 Digital Water as the only option

The ecosystem of stakeholders involved with and impacted by water and wastewater utilities is growing and evolving, spanning across industries, academic institutions, technology providers, and a myriad of other public and private sector players. No stakeholder will be left untouched by the digital transformation of the water and wastewater sector, and all will share the responsibility to step up to the challenges of the sector and secure our water resources for future generations. Most importantly, it will remain the responsibility of water and wastewater utilities alike to ensure affordable access to reliable, quality services to their customers, no matter the challenges the utility may face.

Water and wastewater utilities maintain the responsibility of providing services critical for human health and the well-being of society to communities around the world, services that encompass basic human rights – the rights to water and sanitation. Upholding those human rights will require utilities to embrace a leadership role and forge ahead into a more connected, digital future. Utilities must explore new possibilities and solutions, and branch away from traditional, legacy infrastructure in order to continue providing adequate services for meeting the demands of society. With this book offering both instructions and advice, it is no longer a question of how to become a digital utility, rather, who will be the first to join their utility peers in rising to the challenges and opportunities of the 21st century and beyond. For all the power harnessed by digital technology, no water innovation holds more latent potential than the open human mind.

Water and wastewater utilities must embrace digital solutions. There is really no alternative.

Chapter 2

Instrumentation to data generation

Introduction

The road to digital transformation of the water industry is getting closer and closer but the right decisions must be made at every step of its implementation to make the most of all its exciting potential. The more quality information and prior experience we have on each stage of the digital transformation, the better decisions can be made. Experience in the key points of the necessary instrumentation, the meta-data that should be annexed, data generation knowledge about the uncertainties associated with data acquisition, processing, identification, and intelligent use of the information obtained can be crucial to building a reliable system that can provide maximum utility for each application.

One of the fundamentals to Digital Transformation in the Water Industry is gathering insights from the data collected. An example of this is the smart metering of the water used everyday in households. If the UK had a 100% coverage of smart meters recording on an hourly basis (according to industry standard), over 600 million pieces of data would be collected every single day. Within this data, there are valuable insights which require automated information generation.

Water meter data is only one type of the data collected by the water industry. To obtain value from this data, we need to know three things: what is its purpose and how is it collected; the uncertainty associated with the data itself; and the information that sits behind that data, i.e., the meta-data, providing significance to the raw data. With this information, we can organise the data appropriately, so insight can be extracted based on the following:

- Operations; for an informed position of what is happening in the field, either by manual interpretation or automatically feeding an operational Digital Twin.

- Tactics to inform asset management position and investment priorities to enable water companies to understand how their assets are performing and what needs to be done to ensure that we can protect the environment.

This chapter covers three areas on how the data collected in the industry is transformed into useable information for best value:

- Firstly, we will look at the instrumentation life-cycle and how to strategise for daily data collection. In this way the instrumentation installed in the field measures correctly and is maintained in the right way.

- Secondly, we will look at what is meta-data and how it can help the industry to structure data.

- Lastly, we will look the uncertainty principles, i.e., the importance of knowing what we don't know, and how this is important to consider when we look at our data and translate it into informational insight.

2.1 Instrumentation in digital transformation

The concept of Digital Transformation, especially in the water industry, includes a broad category of techniques and methods that can be used to allow the industry to operate with more efficiency and to make decisions based upon an informed way of working. From the use of modelling techniques to the use of concepts such as Digital Twins, there are common factors that underpin every aspect of Digital Transformation, i.e. the use of data and the source of that data (usually, but not exclusively, some form of instrument that is recording this data).

In terms of Digital Transformation, instrumentation can be defined as: "A set of techniques to measure, transmit, record and regulate physical or chemical variables. Instrumentation is needed to acquire information of a process variable, to be able to react manually or automatically (control actions) to maintain the process in a desired operational point." This includes instrumentation for acquiring information but also valves and controllers for process control.

With reference to instrumentation and its role in the digitalisation of the water sector, key areas of this section include:

- The importance of instrumentation and its role in Digital Transformation;
- The resistance to the effective use of instrumentation;
- The instrumentation life cycle and how it can be used to realise the benefits of both instrumentation and Digital Transformation;
- Some examples of actual case studies where simple and more complex uses of instrumentation can be used to realise immediate benefits.

The role of instrumentation in digital transformation

Instrumentation is a fundamental part of the Digital Transformation of the water industry as it is the principal source of digital data. Instrumentation is present throughout water and wastewater systems and ranges from the use of smart meters at a customer's premises to industrial instrumentation on the various network and treatment work systems within the water industry.

All the examples of where Digital Transformation has succeeded in the water industry so far have been based upon three basic tenets:

(1) good quality data from properly installed instrumentation;

(2) a basic knowledge of the uncertainty of the data; and

(3) robust instrumentation maintenance processes, making sure that instrumentation accuracy is maintained. Conversely, it has been poor quality data, from either poorly installed or poorly maintained instruments, that has resulted in the failure of some of the most promising Digital Transformation projects.

For the success of the projects, a strategy of how to collect information is needed. This strategy can be in a specific area, such as non-revenue water, or in a more generalised company-based operational area.

An example of this is in the Global Omnium Digital Twin model built for the City of Valencia (Conjeos, 2020). This application-specific Digital Transformation project saw instrumentation installed along with dual redundancy on telemetry outstations, coupled with an understanding of the accuracy of the instrumentation using general uncertainty principles. This has allowed the construction of a hydraulic digital twin that enables operators to not only understand the system performance but also to predict future outcomes. Such functionality can only be achieved using accurate instrumentation which is ideally coupled with the instrumentation meta-data to provide full functionality of both visualisation and analytics.

Clearly, with the right instrumentation, situational awareness of the system can be achieved. Thus facilitating informed decision-making, which is where the value exists for companies within the water industry. As an industry, we know that accurate instrumentation is an absolute must but does not always exist. Why not? Is this due to resistance to the effective use of instrumentation?

Resistance to the effective use of instrumentation

Resistance to the effective use of instrumentation usually starts when instruments are not installed correctly or have been installed for little or no purpose. In these circumstances, there can be a perception that an instrument is not correct which, in turn, leads to lack of maintenance of the instrument and, therefore, additional wrong measurements. This leads to a vicious cycle where the instrument provides inaccurate or useless data (i.e., useless information) and it is consequently abandoned. The risk in this approach lies in the use of incorrect data which, in some cases, can cause poor control of the treatment works and result in regulatory issues. The root cause of a lack of trust in instrumentation is due to the following:

Instrument reliability: There is resistance to the use of instrumentation to full effectiveness as it is perceived as unreliable. This can be true if an instrument was badly installed or installed in the wrong place. However, in other cases, the instrument reliability is compromised by poor maintenance.

The threat of instruments: There may be a perceived threat that instrumentation and automation will be used to retrench or replace the workforce. On the contrary, instrumentation should be a tool for operators to operate more efficiently by reducing the time spent manually analysing samples.

Over-design of the automation system: Over-design and then use of instrumentation means the system is over-complicated and inoperable. This causes a gap between the design engineer and the user.

Poor use of current data and poor data management: Instrumentation that is currently in place at treatment works normally feeds through to a SCADA (Supervisory Control And Data Acquisition) system. However, the vast majority of data that the instruments produce is generally not used, leading to "data richness but information poverty".

A lack of understanding of what instrumentation can achieve: There is generally poor knowledge over what instrumentation can achieve to deliver basic and advanced process control.

Poor integration of the current instrumentation leads to the loss of most of data and information produced, causing poor efficiencies in current process control and the inability to utilise the instrumentation to its full effectiveness.

Lack of trust in instrumentation: Instrumentation is not trusted from the operator level to the corporate level or at the regulatory level therefore, it could be purposely not used due to lack of trust.

Below are some examples where some of these factors have influenced problems in the use of instrumentation. Figure 2.1 shows a typical wastewater flow measurement system being monitored for regulatory purposes using an electro-magnetic flow meter. The flow meter is however poorly installed.

An accurate flow meter could be used:

(1) as a pollution warning monitor in ensuring expected flows are received by the works;

(2) to control the secondary treatment process further increasing the efficiency of the treatment works;

(3) to portray a true situation awareness of the operation of the works treatment systems and its performance; and

(4) for the long-term information for asset planning purposes as to whether the data indicates growth or infiltration in the wastewater network.

However, due to poor installation of the flow meter, none of the above can be achieved in this case. In fact, the flow meter is directly in front of an actuated control valve and so the accuracy of the data is being compromised. The flow meter is being used to control the pass forward flow which, because of the error of measurement, is causing a semi-permanent discharge to the site storm tanks and potentially affecting the environment due to poor flow measurement and poor flow control, hiding the true performance of the treatment works. This is the direct consequence of the data's poor accuracy.

It is not only poor installation that can influence the resistance to the effective use of instrumentation. Figure 2.2 shows a large flume which was measuring inaccurately. The lack of maintenance, due to poor access to the flumes caused by the use of heavy covers, prevented effective maintenance resulting in poor performance of the flow measurement system. This had a similar effect to the example in Figure 2.1. In this case, however, the lack of maintenance, and therefore the inaccurate measurement, gave the impression that the treatment work was non-compliant to the regulatory limits. This was caused by a build-up of grit that cause a false high reading of the level-based flow technique. This triggered the false appearance of the need for capital investment "due to growth". Once

Figure 2.1 Poorly installed flow to full treatment flow meter

Figure 2.2 A large flume influenced by poor maintenance procedures allowing grit build up

Figure 2.3 Poor installation can breed poor maintenance. This control valve, installed underneath a raised tank, is not accessible for maintenance purposes

properly maintained, the resulting flow figures decreased by 22%. Poor maintenance, as in Figure 2.2, can be exacerbated by poor installation as seen in Figure 2.3. Instrumentation and automation equipment need to be accessible for maintenance purposes. If there are accessibility issues for practical (i.e., the flow meter cannot be reached easily) and/or health and safety reasons, then the equipment simply will not get maintained and will eventually stop working.

The case studies presented above present barriers to the performance of the instrumentation systems because of (1) poor installation, (2) physical barriers where health and safety was not considered in the initial installation, and/or (3) poor operation and maintenance of the instrumentation assets. Such barriers lead to poor confidence in adoption of the systems that will generate the data to support digital transformation. The experience of these barriers has led to the development of the instrumentation life-cycle philosophy.

The instrumentation life cycle

The instrumentation life cycle has five stages. These are intended to take the designer and operator of instrumentation through the operational life of an instrument and highlight early on issues that could cause problems in the future. The first three stages of instrumentation life-cycle assessment are to help users think about the process of instrumentation and understand the value that an instrument brings. The five stages are illustrated in Figure 2.4 and detailed in the sections below.

STAGE 1 - INSTRUMENTATION PURPOSE

The first stage defines the purpose of an instrument in the water or wastewater system, the data it will produce and how this is going to satisfy an information strategy, thus addressing and clarifying the real application of the instrument. The reason why an instrument is needed could be multiple, including (but not limited to):

Regulatory

An instrument is sometimes required for legal purposes, under the operational legislation (for instance to ensure compliance with an environmental permit);

Financial

An instrument is required for financial purposes (for example for the billing of an industrial customer);

Monitoring/Alert purposes only

Some instruments could be in place for monitoring or alert purposes only. For example, a rotation sensor on a trickling filter or a settlement tank that will inform about a lack of rotation;

Asset Monitoring or Protection

Some sensors are installed on critical assets to ensure that their condition is monitored. The asset is therefore protected, and its state monitored to force the asset to stop in the event of a problem to prevent any damage. For example, this could be a vibration sensor on a centrifuge which can either report on its condition or stop it if there is undue vibration;

Control Purpose

An instrument or instrumentation system that is required to report upon the process variable for the control system to operate around a set point. For instance, a dissolved oxygen monitor opening and closing a control valve on an activated sludge plant to regulate the dissolved oxygen concentration.

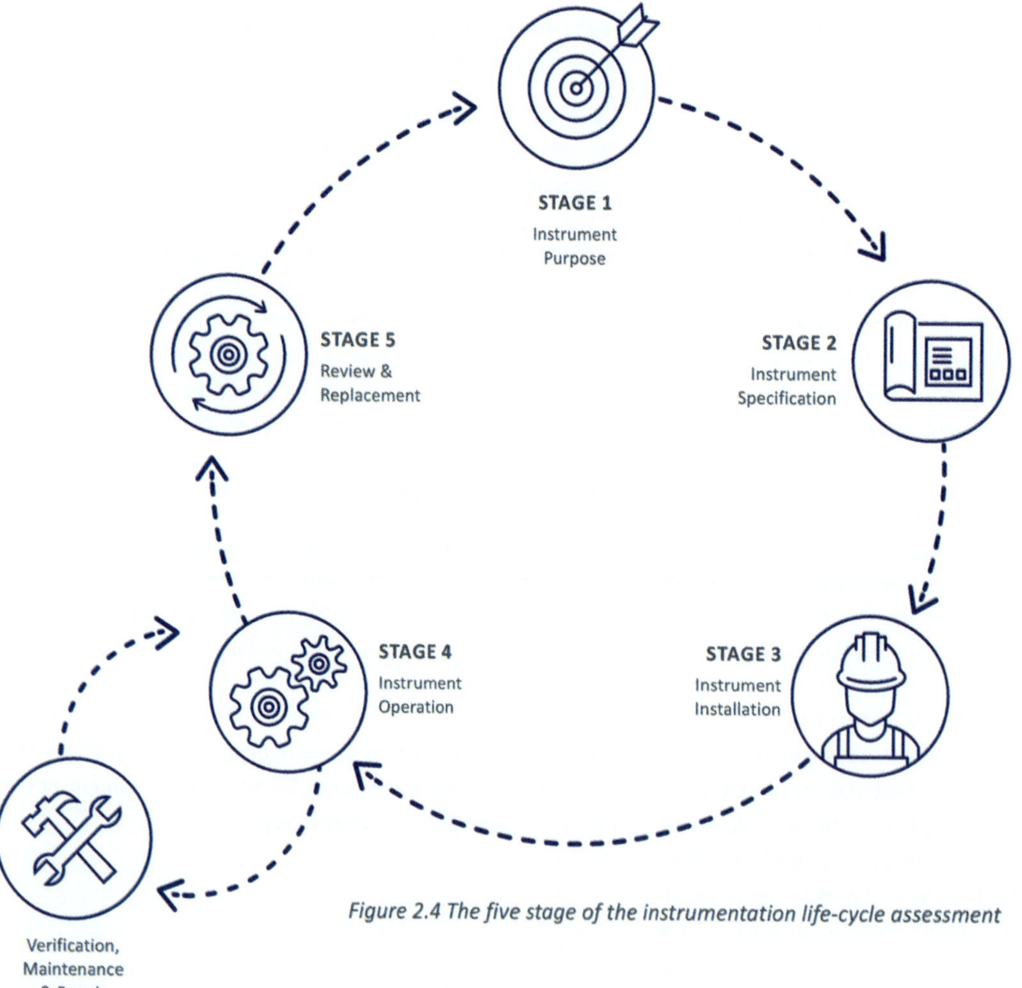

Figure 2.4 The five stage of the instrumentation life-cycle assessment

STAGE 2 – INSTRUMENT SPECIFICATION

The second stage is the instrumentation specification and selection. For this, it is important to understand:

- What parameter is the instrument meant to measure (level, flow, temperature, state)?
- How is it meant to measure it? What technique is going to be used? What is the accuracy requirement? In what range it needs to operate? What is the required response time and measurement frequency?
- What is the application (e.g., in the network, on the inlet or outlet of the treatment works)?
- What are the physical constraints of the measurement location?
- What are the power and communication requirements?
- How is the instrument going to be operated and maintained?
- What are the sample conditioning requirements such as sample delivery, filtration and sample preparation and how? Is this going to affect the measurement?
- What are the costs involved in the purchase and the operation of the instrument (e.g., ongoing chemical cost and)? Or ongoing consumable costs?
- What are the legal limitations of installing the instrument? If some legal schemes are in place, the instrument may have to reflect this limitation.

The examples in this list, albeit not exhaustive, can have a significant impact on whether and how an instrument is installed.

STAGE 3 – INSTRUMENT INSTALLATION

The third stage is to consider the instrument installation and how this is going to be achieved including ability to access, verify, calibrate, maintain, and replace the instrument itself. This is an iterative process, as an instrument may be ideal in terms of specification but not installation requirements.

Before installing any instrument, it is advisable that to be fully aware of the limitations of the technology as well as local conditions that may affect these limitations. These might be physical (e.g., a bend in a pipe or a control valve), chemical (an interfering substance within the water) or potentially biological (algal growth). Understanding these interferences and how they can change through time is essential to ensuring long-term instrument accuracy and reliability.

At this stage, it is also vitally important to understand how the instrument is going to be maintained and eventually replaced as a lack of maintenance (due to inappropriate installation) can affect the instrument's reliability. At the end of the instrumentation asset life, replacement will result in significant disruptions and cost implications. On the other hand, if future replacement is considered prior to installation, in the long run the cost of the instrument and its replacement can be significantly less.

Therefore, putting together an operational maintenance and instrumentation replacement plan is worth the investment in time.

STAGE 4 – OPERATION

The fourth stage is the operation and maintenance of an instrumentation system. This should include an operation and maintenance plan based upon the manufacturer's guidelines and adapted based on practical evidence including:

- Instrumentation cleaning frequency and methodology of how to achieve proper cleaning;
- Instrumentation end-to-end testing;
- Instrumentation calibration versus instrumentation primary verification;
- Instrumentation secondary verification techniques;
- Instrumentation consumables (chemicals, wipers, etc.).

These will vary depending on the instrument type and location. There will be some fixed maintenance (such as chemical replacement) that must happen for the instrument to function, as well as occasional maintenance depending on performance. For example, electro-magnetic flow meter on a gravity inlet with chemical dosing in front of it requires significantly more maintenance than an electro-magnetic flow meter on a pumped final effluent line. The operation and maintenance phases are circular during the life of the asset and can be measured using primary and secondary verification to predict when an asset is likely to fail.

STAGE 5 – REVIEW & REPLACE

The fifth stage begins as the instrument is about to fail and comprises the review of its lifespan, its usefulness and whether and how it is replaced. The first part of the fifth stage is to review whether the instrument has achieved what was decided in the first stage and whether a replacement instrument is required. The purpose of the instrument may well dictate the outcome of this review. If the instrument is required for regulatory purposes, then replacement is needed, whereas, if it is being used for redundant monitoring then its decommissioning is a better option. The main question to ask in this process is whether the replacement of the instrument is viable. If so, the instrument life cycle starts all over again considering that the first stage has already been completed. Re-assessment of the second stage is needed to account for any technological improvements that have taken place in the life of the previous instrument. If the instrument has not achieved its purpose, its decommissioning should proceed.

CALIBRATION

Is the verification and adjustment of an instrumentation system against a traceable reference standard?

PRIMARY VERIFICATION

This is to verify that an instrument is setup correctly and the instrument is working within design.

SECONDARY VERIFICATION

This is the test that compares an onsite monitor to a traceable standard to ensure the measurement system is measuring correctly.

END TO END TESTING

This is a test to check if the transmission of the instrument signal is correct at the instrument end and the telemetry end. Faults are normally due to scaling issues.

It is important to decommission an instrument as an abandoned instrument has the potential to lead to a lack of trust in instrumentation.

In summary, the instrument life cycle is a tool used to ensure the accuracy of instrumentation. This is absolutely vital within the Digital Transformation concept as the majority of projects have failed due to poor quality data. In fact, for instrumentation to be fully utilised, it is vital to consider the instrumentation meta-data such as the instrument location, its purpose and uncertainty and how these will affect the decisions that are made in terms of regulation, control and asset management. These all contribute to the value of data and how it can be used to increase the effective use of instrumentation.

The effective use of instrumentation in digital transformation

What are the benefits of getting instrumentation right? Below are a few case studies of the effective use of instrumentation and how it can be used in Digital Transformation.

A SIMPLE VIEW OF THE EFFECTIVE USE OF INSTRUMENTATION IN DIGITAL TRANSFORMATION

The first case (Figure 2.5) shows the flow data from a wastewater treatment works that has been converted to a total daily volume taken over several years. The instrument is a simple 100mm electro-magnetic flow meter that has been used for monitoring and regulatory purposes. Although the meter was not installed as part of a Digital Transformation project, the data produced can be used to improve performance.

Figure 2.5 shows four clear peaks which grow over a period of three years showing the gradually worsening infiltration into the sewer environment. With additional information, such as rainfall, geology, and the performance of the system, this can be used to predict where the source of the problems within the sewer network is. This is a long-term asset planning scenario using total daily volumes, but even at a short- term operational time-scale, valuable information can be discerned from a simple flow meter. These include the performance of the flow control system and the works control of flow to full treatment or potential blockages within the wastewater system (collection network or the wastewater treatment works itself due to low flow events).

A MORE COMPLEX VIEW OF THE EFFECTIVE USE OF INSTRUMENTATION IN DIGITAL TRANSFORMATION

The common idea connected to Digital Transformation is that this is based upon a more complex model-based system fed by instrumentation working in real-time (similar to the Digital Twin approach). This is being seen in both the wastewater and potable sides of the water industry where the approach is being used for detection, and in some cases, control of non-revenue water.

In water distribution systems, the Digital Transformation approach has been in place for many decades. For instance, distribution systems have been using a systematic approach in creating a district metered area (DMA) and the subsequent flow monitoring of inputs and outputs to understand where the water is going. The big leap that water companies have taken is then using this data to drive their asset management programmes instead of replacing the oldest assets. An example of this is from Lisbon (amongst many other cities), where the operator used this approach to reduce the non-revenue water from 23.5% to 7.8% over an 8-year period, from 2005 to 2013 (Sardinha et al. 2017). The local water company for Lisbon, EPAL, used the flow data to identify water loss locations and then target asset replacement, resulting in a significant decrease in water loss from the system (Sardinha et al. 2017). There are many examples of reducing water loss and managing assets with more complexity such as using a digital twin, a model-based system complete with instrumentation inputs.

The wastewater collection system has also a wide variety of innovations that could be included under the general title of Digital Transformation. In the water network, these innovations mainly refer to the control of the hydraulic aspect of the system as this is the simplest component to control and has the highest impact on the customer and the environment. Similar to water distribution systems, this actually started to happen long before the current trend of "Digital Transformation," with smart wastewater collection networks first being installed in

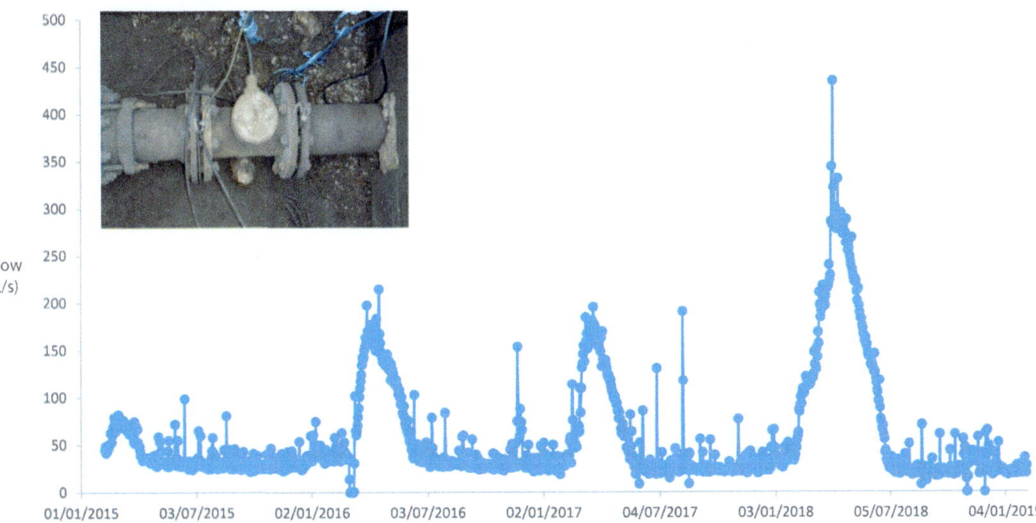

Figure 2.5 Flow data from a rural treatment works

Minneapolis in the 1980s along with systems in Europe, including the one that was developed for the Barcelona Olympics in 1992 (Kellagher & Osbourne, 2013). These systems have become more and more advanced but generally have worked with the basis of a model with inputs from weather service data and sewer level monitors. More developed countries have installed instrumentation within the network and have been controlling flows. However, the general reliability of these instruments within the sewer environment as well as the complexity of installation have been a barrier because of the harsh conditions within the sewer network, the difficulty of installation due to poor accessibility, the difficulty of installation due to a lack of power and telemetry sources as well as the difficulty of instrumentation maintenance.

There are also examples of technologies being successfully used within the wastewater network from basic to data analysis which has been later integrated into decision-making. A prime example of this is the Eastney Project (Ellison, 2016) in Southern Water in the UK that used a combination of a wastewater collection network along with sewer level monitors, rain gauges, and weather radar to predict the impact of flows into the wastewater system. The main driver for this project was to protect the city of Portsmouth from pluvial flooding of the sewer network following an incident in September 2000.

This example shows that the use of digital tools does not need to be complex. The global water industry is recognising this and as companies build their data and information strategy that maximises the value from the instrumentation they already have. In this case, Digital Transformation is not about installing more instruments, it is about getting the value out of the existing data sources, rationalising where necessary, and only installing more instruments where there is a value identified.

Instrumentation is a vital part of the Digital Transformation of the water industry. To overcome resistance to effective use, it must be as accurate as possible and uncertainty in measurements must be known. Fundamentally, it is important for companies to realise the value of instrumentation through the actual use of data and the corresponding information to drive (a) situation awareness and (b) informed decision-making. This can be achieved through application in the short term, but the full value is in setting a data and information strategy so that the instrumentation needs of the water operator can be identified. The instrumentation life cycle can identify the best suited instrumentation and the value of data and information it produces.

Next, we will look at the meta-data and how this helps identify what the data is, what it refers to, and how this is very important in allowing the user to use data effectively.

2.2 The value of meta-data for water resource recovery facilities

As water resource recovery facilities (WRRFs) enter the era of big data, they are naturally confronted with the challenges of integrating smart actuators, sensors, and autonomous control systems in a sensible and transparent manner. One aspect that remains an important burden to bear by water utilities is the storage and management of sensor data in view of later use. To enable data interpretation beyond the original time of data collection, it is crucial that the collected sensor data is augmented with an adequate description, i.e., meta-data. Indeed, data collected today are expected to be useful in the future to respond to even more complex operational challenges and new demands for the environmental impacts, the produced effluent quality, and resource efficiency. Given that those future challenges are unknown, it is particularly challenging to define the required meta-data for a generation of future-proof data mines, as opposed to data graveyards, and ensure its collection in a timely manner. In this section, we highlight the most important aspects of this meta-data challenge, and we provide arguments and early solutions leading towards harmonised data collection and interpretation.

The need for meta-data

In recent decades, the water sector has undergone an instrumentation revolution. For example, the measuring of dissolved oxygen was introduced first to WRRFs to augment the existing capabilities of flow and level sensors in the 1980s. Since then, the diversity of available sensors has steadily increased. The challenges and opportunities in collecting 'big data' are often categorised into the following four 4 Vs: Velocity, Volume, Variety, and Veracity. Thanks to increasingly efficient communication techniques and extreme reductions in data storage costs, data collection has become extremely scalable. This means that today's WRRFs are now mastering the first two of the four Vs, Velocity and Volume.

Recent developments in data mining, machine learning, and optimisation, enabled by virtually endless computational power and algorithms for computer-based learning, have been met with enthusiasm in the water sector. Many are enticed by new capabilities of computer-aided decision-making both at an operational and managerial level. However, many attempts in advancing automation from the increasingly large data streams invite a hard confrontation with the other two Vs of big data: Variety and Veracity. Human intelligence and smart routines are still needed to categorise, structure, homogenise, and convert data into valuable information. Indeed, this important step easily demands 40% of the costs in most consultancy and data science projects, both in the wastewater treatment sector and others. This cost is largely associated with the need to triage the available data (i.e., separate garbage data from data fit for purpose) to avoid the commonplace 'garbage in, garbage out' problem, which is now more obvious than ever.

The authors of this report believe the cost of this task can be reduced drastically if routine data collection and management practices are updated to support data-intensive decision-making and automation. More specifically, existing data should be augmented by providing information on the original purpose, the data-generating devices, the quality, and the context of these data. This kind of descriptive information is known as meta-data and is an essential ingredient to turn large volumes of raw data into actionable information. Indeed, detailed knowledge about the measurements are needed for sound and creative data analysis, so as to guarantee an impact on design and operational decisions. Unfortunately, there are no wastewater-specific guidelines available to the production, selection, prioritisation, and management of meta-data.

Garbage in, garbage out — Why data analytical tools are picky

There are several reasons why simply collecting more and more data is not sufficient. This holds both when mechanistic models (e.g., physics-based) or empirical models (e.g., including machine learning and other data-based models) are used for data interpretation. Here are a few reasons and required adjustments to resolve them:

(1) Typical data include measurements of suspect quality. It is common to observe the symptoms of short-lived sensor faults in the form of outliers, spikes, high-noise features, and sustained deviations from the reference measurements. Moreover, water quality sensor signals are prone to sustained faults, often due to calibration errors and drift. Including all data without elimination or correction of low-quality measurements will lead to faulty models. A key requirement is that the data used for model identification should be of high quality, either through proper management of the data collection system or through a well-established data refinement process with offline or online data validation and reconciliation tools.

(2) Often, the data available for model identification does not correspond to the conditions under which the model will be deployed. Consider the flow and influent composition data from a WRRF under dry and wet weather. These data should be clearly separated to obtain a reliable model for the typical influent of a WRRF under these distinct operational conditions. Therefore, another requirement is that the data used for model identification should be representative for the deployment conditions.

(3) While data might be voluminous, the patterns one wants to analyse are often rare (e.g., toxic spills, rain events). Indeed, one of the many promises of machine learning is that it can help detecting and diagnosing rare events. To make this simpler, a sufficiently large data record corresponding to the events of interest should be available or any imbalance in the frequency of events should be accounted for through the provision of detailed system knowledge.

Today, the requirements mentioned above lead to a necessary but cumbersome triage before data analysis can take off. Human experts crawl through the large volumes of data and modify,

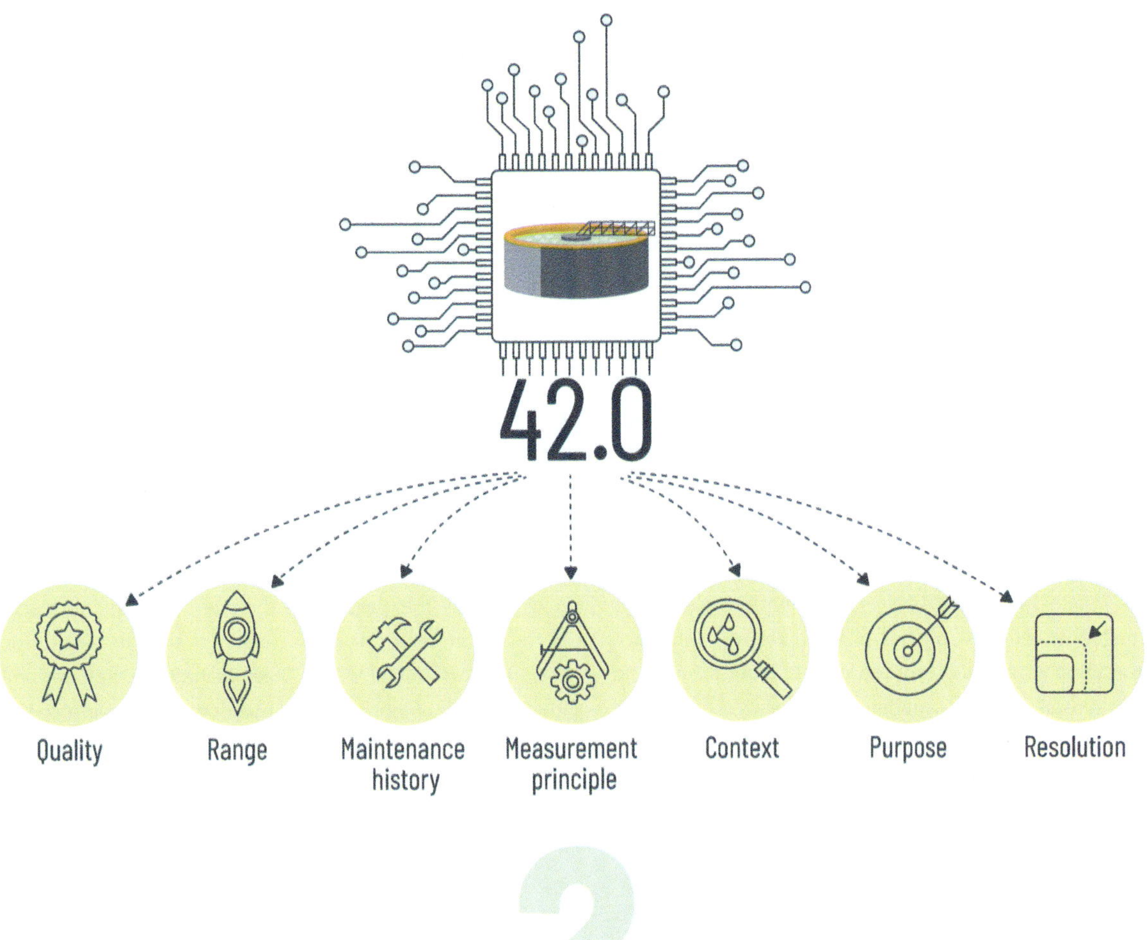

Figure 2.6 Can we interpret the provided measurement equal to 42.0? We need to know a lot more to evaluate the information contained in sensor signals. The kind of descriptive data we need for interpretation is known as "meta-data" and includes the purpose of measurement, the measurement principle, the temporal and measurement resolution, sensor maintenance history, indicators of signal quality, and the spatial and temporal context of the measurement.

select, and annotate the data to enable a correct execution of the computer-aided prediction or optimisation task. This is a tedious effort and often includes subjective assessments by a domain expert. Meta-data, if available, can help. First, informative meta-data can assist by automating the triage to a high degree. Second, structured meta-data reduces the need for subjective assessment, in turn increasing trust in algorithmic predictions and decision-making. Since any measurement-based algorithm relies on representative, reliable, and interpretable data, data sets should be judged by virtue of the provided meta-data, next to the conventional measures of sensor signal quality, such as trueness, precision, and response time (see for definitions). Figure 2.6 illustrates how measurement values alone are not sufficient to reap the benefits of intensive data collection systems.

For data triage purposes, a good data set will include meta-data, about the following aspects:

(1) DATA-GENERATING SYSTEM – Information describing every step of the data collection process, including information about: (a) the purpose of data collection, (b) the sensor hardware (e.g., measurement and temporal resolution, measurement unit, measurement principle, manufacturer, sensor model, etc.), (c) signal management, including recording, transmission, and storage, and (d) data refinement, including all modifications of the data after data collection. This kind of meta-data enables to select data that are fit-for-purpose and is often already available from the sensor devices themselves through a modern digital communication system (e.g., Ethernet).

(2) DATA QUALITY – Detailed information about the sensor signal quality and is based on: (a) a digital record of all sensor calibration, validation, and verification events for every sensor, (b) description of individual records that are suspect (e.g., outliers, spikes), and (c) descriptions of sustained periods with poor data quality (e.g., due to calibration errors, lack of maintenance, drift, and other malfunctions). Such descriptions can be obtained through manual data annotation by a domain expert but, importantly, also by careful deployment of data-analytical tools, in turn leading to quantitative data quality assessment and quality control. This means that also (d) the output from the methods used to collect this information (e.g., algorithmic analysis, standard operation protocols, expert annotation etc.) should be included as meta-data. This kind of meta-data enables identification of data that satisfies the required data quality.

(3) CONTEXTUAL INFORMATION – Information describing the circumstances outside of the plant that may influence the interpretation of signals recorded on the plant. This could be information on: process mode, local weather, including seasonal changes or stormy weather, or meaningful changes in the structure and operation of upstream infrastructures (e.g., sewers). It also includes information on rare events, including operating failures, social gatherings, or toxic spills. Quite often, this kind of information is provided by technical and operational staff. This type of meta-data allows selection of data that is relevant to the task at hand, i.e., that is informative and relevant.

Unfortunately, the meta-data described above are rarely available. For example, descriptions of the procedures (e.g., sensor maintenance) might be missing, results from sensor validation may not be logged, and the circumstances under which data were collected (e.g., stormy weather) could be unknown. This kind of information is indispensable, however, for data-intensive prediction and optimisation of the performance of WRRFs. Without it, one critically relies on the memory of on-site staff to interpret the available data. Within a year, historical data can become useless for most data-based tasks as personal memory fades and the collected data meet their expiry date. This loss of valuable descriptive information can quickly turn information-rich measurements into a data graveyard. Furthermore, this lack of descriptive information often only becomes apparent a long time after the original measurements were collected. Thus, effective data governance not only requires that one can answer today's questions based on data but also to manage the collected data in such a way that future yet unknown questions will be answered reliably too. To account for these unknowns, as well as for an ageing work force, to empower staff members, and to assure the long-term utility of historical data records (e.g., decades) for important decisions at operational and managerial level, collection of meta-data of the types discussed above should become a routine matter.

Structuring meta-data

Even when meta-data is safeguarded for later use, it may be challenging to wield it. Indeed, meta-data frequently resides in a vast array of design specifications, manuals, protocols, and spreadsheets, often stored in separately managed databases and folders (i.e., silos). The location of, and access to these data is often managed in an ad hoc manner. To enable the envisioned triage of data with a highly automated process, meta-data should be accessible and stored in a way which allows for an easy navigation. To achieve this, meta-data needs to be stored in a structured manner and routinely kept up to date. Where feasible, a centralised system for meta-data storage will be helpful to manage access while ensuring accuracy and completeness.

The definition and integration of meta-data is often a hurdle and typically not part of off-the-shelf software products used in the water sector. For this reason, it is important to focus on ways this can be achieved. These include the generation of the identification of a primary data source, which points to the most accurate and complete version of all data and meta-data. This primary data acts as a single-source-of-truth and engenders a shared and unambiguous understanding of the most current data and information available to all data users. Naturally, identifying a primary data source implies a well-calibrated appreciation of the need for good governance of data, information, models, and software. In turn, this means that these changes affect almost everyone in the organisation, thus requiring a careful alignment of objectives and needs as part of good digitalisation practice.

Sensor maintenance revisited

Many utilities have protocols in place to check the validity of sensor signals on a regular basis. More often than not, a reference measurement is used to determine whether a maintenance action, such as intensive cleaning or calibration, is needed. However, this kind of reference measurement is rarely recorded. To illustrate how a modest enhancement of existing quality assessment and control practices can lead to useful meta-data, we take two series of reference measurements. In Figure 2.7, one can see the offset of two pH sensors as a function of time during their 2-year deployment.

This offset is the measured potential of the sensor in a calibration solution with pH 7 and is available as part of the sensor calibration curve stored within a typical transducer. One sensor (sensor A) exhibits a monotone, decreasing profile. This decrease is easily explained by drift of the reference electrode. This drift is compensated in practice by regular calibration. In contrast, the second sensor (sensor B) exhibits an increase of the offset after 500 days. This is due to irreversible wear-and-tear of the sensor and cannot be corrected with calibration. Importantly, making a distinction between the offset around

180 days (-11.4mV), which is explained by normal drift, and the similar offset value at 510 days (-11.5mV), explained as a result of damage, is only feasible thanks to the whole history of offset values. Without these meta-data, such a diagnosis is infeasible.

We therefore encourage the systematic recording of this kind of meta- data before and after every maintenance action already executed on the plant, including cleaning, calibration, and part replacements. This will enable an accurate and timely response to sensor wear-and- tear, thus improving data quality. In the long run, it can also produce valuable information to implement preventive sensor maintenance and decide on the best sensing hardware, particularly by quantifying the trade-offs between the cost of sensor hardware against the obtained data quality and costs of maintenance actions.

The meta-data workforce

As mentioned above, effective digitalisation requires cultivation of good meta-data management practices, many of which can be automated through careful selection and structured management of meta-data.

This has produced a range of novel roles in the wastewater sector, with names like data steward, data engineer, or chief data officer, highlighting the need for in-house expertise in data handling, managing of expectations regarding digital transformation, and translation of opaque computational concepts into a common-sense language. These experts can be extremely helpful to turn stale databases into an effective source of information for operations and investment decisions. They should be tightly integrated into the existing work force, to facilitate early adoption of evidence-based decision-making and to ensure alignment of expectations and objectives across the organisation. In the future, we hope specialised teams and staff can reach across the following topics of relevance, which are often handled by different subject matter experts today, each with their own isolated terminology:

1. Add and integrate new devices into an existing control system.

2. Manage and optimize data collection systems in multi-purpose settings, e.g., data for real-time control, for reporting, for model construction and validation and for planning major upgrades.

3. Manage and optimise sensor data quality aided with basic and advanced data validation tools.

4. Augment existing data streams with meta-data, as described in a highly automated fashion.

5. Merge the meta-data needs for WRRF operations, algorithmic requirements, and current sensor data collection routines.

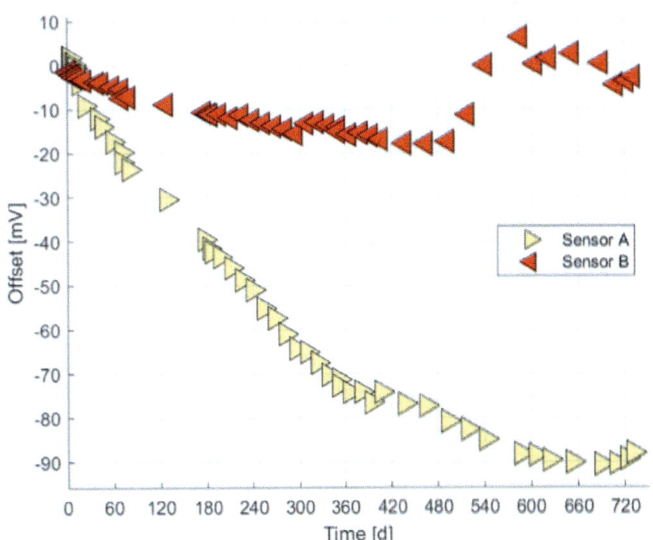

Figure 2.7 Effects of wear-and-tear on two pH sensors. The sensor offset (electrode potential at pH=7) diminishes gradually as time progresses for sensor A. This is attributed to drift of the reference electrode and can be accounted for by calibration. The accumulated drift is about - 90 mV by the end of a two-year period, which amounts to a shift of roughly 1.5 pH units in absence of calibration. In sensor B, the offset increases after 500 days of use. This is attributed to irreversible damage to the sensor and requires sensor replacement.

Getting started with meta-data

Following enthusiastic responses to breakthroughs in artificial intelligence, robotics, and machine learning, it is increasingly clear that reaping the benefits of intensive data collection requires augmentation of sensor signals with descriptive information. The need for this descriptive information, called meta-data, and the challenges to obtain and manage it underscore that there is no free lunch. Effective data governance includes the provision of high-quality meta-data and will make the difference between failures and successes in data- intensive system monitoring, automation, and optimisation. As a very first step towards data-wise management, we recommend WRRF managers to:

(1) Initiate the automation of meta-data collection by enabling data integration and meta-data storage of sensor maintenance actions (installation date, calibration, cleaning, validation, and verification).

(2) Assure availability of basic meta-data for online sensor signals in the same location as the sensor signals (e.g., same database). A very basic set of meta-data consists of:
a. Unit of measurement
b. Measurement range
c. Measurement resolution
d. Measurement principle
e. Sensor location

(3) Prepare for advanced meta-data practices, including the provision of complete historical records of:

a. The roles and/or purposes of the sensor

b. History of measured values of sensor offset sensitivity, trueness, precision, and response time

c. History of operational state (operational, calibration, validation, maintenance)

d. History of maintenance protocols for sensor calibration and validation

(4) Evaluate the potential of any type of meta-data to prevent expiration of precious data and turning what are currently data graveyards into valuable resources for decision-making.

Once we have the meta-data in place, we actually need to know how accurate the data is and how "uncertain" we are about the data. This has been a discipline of its own within instrumentation and of course other areas. Typically, this has involved using the General Uncertainty Method (GUM). To know the uncertainty of the data that we are recording is absolutely vital when we move forward and "digitally transform" as the volumes of data that we have are not going to give us a chance to "correct" any errors even if we know what they are. This is the importance of knowing what we do not know which is what we will cover now.

2.3 The importance of knowing what we do not know

Uncertainty is an often unnoticed aspect of everyday life. The progressing digitalisation of the water industry brings more and more data and models into the heart of decision processes. While abundance of data and availability of data analytics tools hold a lot of promise, a wide range of challenges are also associated with different types of uncertainty in data. This section aims to raise awareness of the many sources of uncertainty in digital decision-making and describes how digital water approaches and tools can help us to stay in control and make decisions with (at least some) confidence.

Life's full of uncertainty and this is probably one of the things that makes it interesting. We don't know today's special at our favourite restaurant until we go there to find out. We might get stuck in traffic on the way to work if an accident occurs, but we do not know whether or not this will actually happen. And we do not know whether an asteroid will crash down on Earth and wipe out 90% of vertebrate life in 2 years' time. We may choose to be surprised by the restaurant's chef or have a sneak peek at their website (and choose to go elsewhere if we do not like today's special) but cannot know whether a road accident will occur (if we could, we'd take the train instead). And of course, NASA is looking for potential Earth colliding objects, but if they

find one heading straight towards our planet, what can we do about it? These are just a few of the myriad of examples that could be chosen to illustrate the uncertainty we are either accepting, by choice or inevitability, or working around in our everyday lives, on different time scales.

Uncertainty in urban water services

Urban water service providers are also dealing with a wide range of uncertainties in their operations and planning. Extreme weather events, which are rare and unpredictable at all but the shortest time scales (but becoming more severe and frequent), pose problems for storm water systems. In many parts of the world, drinking water is sourced from rain-fed rivers, that may be unpredictable (again on all but the shortest time scales) in terms of their discharge and therefore a river's capacity to supply sufficient water to meet our needs. Drinking water demand, on the other hand, is easier to predict on short time scales (that is to say, it is on the level of a city, but not on the level of an individual person) than in the long term, because behaviour changes, new types of water use appliances come to the market, people move, etc.

Why does uncertainty matter?

In a sense, urban water service providers, or at least most of them, are used to long-term thinking. The urban water infrastructure has a life cycle of at least decades but more often longer than a century. This means that thought needs to be given to the requirements not just of today, but also of the decades ahead. And this requires thinking about (uncertain) changes in demand, population size, climate, etc. But there is another reason why uncertainty matters for urban water services. As we are moving towards digital water systems, more and more decisions (either taken by humans or by systems) are based on data and models. For example, the current trend of developing digital twins (twins of the physical infrastructure prototype) requires the understanding of the uncertainty embedded in them as there is no perfect digital replica of complex infrastructure systems. Thus, a more solid basis for informed decision-making is introduced, but possibly also a misconception of the completeness and reliability of these data and models.

Data may comprise different types of uncertainties (think of representativeness, measurement errors, communication errors), and models even more so, as long as the urban water service providers do not have accurate and complete descriptions of all their assets. To give an example, if you do not know to what degree the capacity of your cast iron drinking water pipe has decreased due to corrosion and tuberculation, it is impossible to calculate how much water is transferred by this pipe without costly measurements. This means that uncertainty in system state information may result in incomplete understanding of the system's behaviour or the wrong decisions being taken (if there even is a clear right or wrong decision). And this is even before considering human behavioural and economic factors, which introduce beliefs, desires, emotions, norms, etc. in addition to rational considerations.

Current developments

The most important of the longer-term developments is climate disruption (Pidcock et al., 2019). It is obvious that this will have significant consequences for water management and the human water cycle. Peak water demand is increasing, the availability of water for the production of drinking water, but also for agriculture, is decreasing, the peak discharge of rainwater is increasing, etc. But many uncertainties remain in the magnitude, timing and local effects (in part because they also depend on our future mitigation actions). As it well known, the warming up of Earth's climate will not remain limited to the current value of about 1 degree since the pre-industrial era (Masson-Delmotte et al. 2018) but will most likely continue to about 2.4 degrees by 2100 (Climate Action Tracker 2019). The temperature rise we have seen so far could, therefore, occur again in just a few decades, i.e., within the lifetime of the infrastructure now being designed and constructed. This brings with it new extremes, although it is difficult to predict their magnitude.

Another relevant development, which is also a source of some uncertainty, is that of population size. Population growth and decline due to births and deaths can be predicted within a reasonable range. However, migration is the most important factor, at least in large parts of Europe and North America. Geopolitics is an unpredictable factor, but climate disruption is possible to an even greater extent. The latter applies both to areas threatened by sea level rise and, even more in the shorter term, to areas that become unlivable due to extreme temperatures and limited availability of water, but also to the areas that will absorb climate refugees. Additional factors related to the population include ongoing urbanisation, changing customer expectations, etc.

These developments introduce uncertainties in the demand for and availability of water, and the amounts of storm water that require dealing with. We need to be aware of these uncertainties when designing and constructing new infrastructure and adapting existing infrastructure.

Different kinds of uncertainty

A distinction can be made between three types of uncertainty of a technical nature, namely unpredictability, structural uncertainty (especially in models), and value uncertainty (especially in data), see Table 2.1 (IPCC, 2007). All three occur in our modelling of water systems and decisions being made. They are associated with the state of objects and the completion of processes, and include stochastic (i.e., randomly occurring) and unknown (i.e., not determined or registered) parameters.

When considering the future, a limited number of scenarios is often used. These describe a potential future event or outcome. Though they have their value, they do not provide a sufficient answer to dealing with uncertainty. It is generally impossible to assign a probability to an individual scenario. Also, a set of scenarios is not intended to be completely comprehensive, nor can it ever be. And it is often difficult to determine which aspects in a scenario are expected to remain the same (stationary) and which are expected to change (non-stationary).

On top of this, there is also 'deep uncertainty' (a particular class of wicked problems) where experts do not know – or cannot agree on – the best way to deal with a problem. For example, the deep uncertainty in projections of future sea level rise. This type of uncertainty makes it impossible to rank different aspects of a problem or different scenarios in terms of importance or probability (deepuncertainty.org 2017). The term 'deep uncertainty' is used in particular with respect to decision processes.

Reliable and resilient water infrastructure for an uncertain future

There are at least three categories of approaches to deal with this uncertainty using technical methods. To prepare our water infrastructure for the uncertain future, all three will have to be applied.

The first approach is to make people and/or water systems perform well despite changes in our environment and/ or conditions. This starts by operating most efficiently at present, i.e., under current conditions. As an example, additional methods for leakage control (ranging from pressure management to model or inspection-based leak detection) can be employed. Much work has already been done in this area and many tools are available. Nevertheless, there is still a world to be gained here, given that high leakage rates still occur in many parts of the world including those with water scarcity. Digital water approaches can be a real game changer here, providing tools to accurately detect and localise leaks.

Also, the reduction of drinking water demand (by behavioural changes and/ or the installation or upgrade of equipment to use less water) (Blokker et al., 2017) helps to get the most out of existing infrastructure and postpones investment in new infrastructure. Easy local reuse options such as rainwater harvesting can be encouraged. Then making sure that the system continues to perform well under changing conditions through infrastructural adaptations, for example by increasing the capacity to transfer water between supply areas to mitigate local water scarcity. Doing this successfully requires digital modelling and optimisation tools and an adequate digital representation of existing infrastructure and its workings.

Secondly, it is important to increase the resilience of our water supply systems (i.e., the ability to respond flexibly to circumstances that may or may not have changed temporarily) (Nikolopoulos et al., 2019). Monitoring, e.g., by sensors and/ or remote sensing, helps to understand circumstances and the water system's response to them. A concrete example of the resilience referred to is increasing the possibility of switching between different source types (Stofberg et al., 2019) and source areas (geographically). In addition to the availability of these sources, this also requires sufficient capacity for treatment, storage, and transport of the water to the right location. This balancing act is of significant complexity, and best

TYPE	EXAMPLES OF SOURCES	TYPICAL WAYS OF HANDLING
UNPREDICTABILITY	Unpredictable expressions of behaviour such as the development of political systems	Use of scenarios with a plausible range with a clear indication of these and assumptions, limitations and subjective judgements
	Chaotic processes in complex systems	Judgements Ensembles of model simulations
STRUCTURAL UNCERTAINTY	Inadequate models, frameworks or structures	Clear specification of assumptions and system definitions
	Ambiguous delimitations or definitions	Comparison of models with observations for a wide range of conditions
	Misrepresentation or neglect of relevant processes or relationships	Assessment of maturity of the underlying science
VALUE UNCERTAINTY	Missing, inaccurate or unrepresentative data	Analysis of statistical properties of sets of values (observations, model results, etc.)
		Hierarchical statistical tests
	Wrong resolution in time or space	Comparison models with observations

Table 2.1 Summary of the Three "Technical" Types of uncertainty according to the IPCC (IPCC 2007)

supported by digital water technologies for design, optimisation and monitoring of resilient water systems. Note that efficiency and resilience are generally competing objectives — an efficient system will have shed the overhead/redundancy required for resilience.

The third approach is to explicitly include uncertainties in model evaluations and predictions. This approach can be combined with the second one in robust/resilient optimisation to maximise the capability of a system to deal with adverse circumstances under uncertainty. By quantifying uncertainty as much as possible and expressing it as part of predictions (whether of future water demand or the performance of the system), we at least would know how much trust we can put in these predictions. Moreover, this makes it possible to search for a design or an operating policy that meets the set requirements with a certain probability, which one finds acceptable.

For example, one can imagine a water supply system that is capable of supplying 90% of water demand even in the 10% driest summers expected over the coming 40 years. And if we cannot estimate probabilities, we can try to find a solution that performs well under most scenarios.

One-off to continuous process

It is tempting to think that by realising a design or an operating policy that was optimised for its capacity to deal with adverse conditions, it is enough for the existing uncertainties to be taken care of. However, not all the uncertainties in, for example, how the environment develops can be captured with this approach. This means that in addition to robustness, adaptability to new circumstances is also an important factor in the design (Kwakkel et al., 2016). Moreover, this means that designs must be scrutinised from time to time and adapted where necessary.

The Dynamic Adaptive Policy Pathways (DAPP) approach provides a framework for this. This approach considers both measures to be taken to deal with circumstances of the near future and measures that keep options open for later implementation to deal with circumstances of the more distant future. It describes a set of partially consecutive policy actions and their interactions, and monitors changes that lead to the failure of immediate measures, requiring the new measures to be put in place. For example, putting buckets under a leaky roof instead of replacing the tiles is a cheap solution in the short run, but this may result in the beams of your roof rotting away, which eliminates the option of just replacing the tiles in the future (since the whole roof will need to be replaced then). But until you manage to replace the tiles, the buckets may be an acceptable temporary solution.

	DESCRIPTION	EXAMPLE APPLICATIONS
(peak) Demand forecast	Longer-term water demand forecasting, based on (a) assumed future behaviour and technology or (b) relationships in recent water demand variations	Blokker et al 2017, De Souza Groppo et al, 2019
Design and optimisation tools	Design, operation and dimensioning of infrastructure, numerical optimisation based on objectives and preconditions	Morley and Savic, 2020, Maiolo et al, 2017, Jiang et al, 2020
Scenario and optioneering tools (which calculate the consequences of scenario choices) and serious games	Tools for calculating (e.g., capacity) the impacts of different scenarios and options for building systems and policies related to these systems	Nikolopoulos et al, 2019, Water Sensitive Cities Scenario Tool
Serious games	Tools for evaluating and visualising the effects of policy decisions in an abstracted but meaningful way in order to facilitate discussions and decision-making processes in a multi-stakeholder setting	Savic et al 2016, SIM4NEXUS, 2020

Table 2.2 Types of tools for dealing with uncertainty when considering or predicting drinking water supply

The digital water paradigm provides the tools for dealing with uncertainty

Hydroinformatics, operating at the interface of ICT and water applications, offers a toolbox to deal with the various aspects of uncertainty. Table 2.2 gives an overview of the types of tools available to support strategic/policy choices and the design of subsystems of the drinking water supply. The successful application of these tools hinges on the availability of adequate digital representations of systems and infrastructure and their behaviour, i.e., models and data. An illustrative selection of example applications is included for demonstration purposes.

Reducing uncertainty

Structural uncertainty (unsatisfactory models, frameworks, or structures in various ways, e.g., the detailed layout of a water network) in physical models can, in principle, be reduced gradually by continuous improvements in model representations. These improvements may arise from insights gained from the practical application of models. This is, however, not true for models that include social and behavioural elements, because of potential feedback between the model and behaviour.

But it is the value uncertainty in particular that provides the best opportunities for improvement. The knowledge of our systems and processes increases rapidly with the introduction of sensors and inspection techniques, which provide us with more information about the processes and also about the infrastructure itself. And it seems that we are only at the beginning of this development. Also, environmental data is increasingly becoming freely available, which allows us to reduce uncertainty by replacing generic estimates or extrapolations from low resolution data previously used in models and decision support systems.

Communication of uncertainty

We have become used to seeing and understanding one of the simplest forms of visualisation of uncertainties in ensemble and bandwidth plots. These are commonly used in communicating weather forecasts or predictions of global warming and sea level rise including the associated uncertainties. They work quite well for single parameters, but for more elaborate systems, this method of visualisation quickly becomes too complex, and a more sophisticated approach is required. Depending on whether the complete system is considered, or one is looking at a minute detail, different levels of aggregation on the uncertainty bounds can be applied (e.g., aggregated variances to full histograms). This also depends on who is looking at the data. Researchers doing data exploration need more control on the visualisation of

uncertainties than end users of data presentation. Augmented Reality (AR) applications may help to reduce uncertainty with respect to, e.g., infrastructure layout and geometry - an overlay of the digital (holographic) representation of the infrastructure on top of the real thing makes deviations very easy to recognise (Figure 2.8). This AR technology is particularly useful for field personnel to see the augmented view of buried water infrastructure and locate possible solutions, e.g., detect valves to be closed to isolate a leak. On the other hand, Virtual Reality (VR) may be a very useful tool to explore in a safe environment various future scenarios, representative of the state of uncertainty, and in optioneering. Future use of VR in the water industry might be to improve visualisation for serious gaming tools.

Are we ready for the uncertain future with Digital Water?

Technically, the representation and propagation of uncertainty need to become standard steps in building and using models to design and operate urban water systems. This applies to models of entire systems, sub-systems (purification, distribution network, storm water network), but also, e.g., to numerical optimisations that make use of these models. Work to include uncertainty in these models has already been done in academic circles, but this has not yet seeped through to operationally applied software in practice.

Applied research institutes and software providers aimed at the water sector still need to develop this operationally applicable software. Also, with regard to methodologies for dealing with uncertainty and decision-making under (deep) uncertainty, a lot of work has already been done in the academic world and in other application fields, which can be adapted for and applied in decisions about our urban water systems.

However, the technical aspects are only one side of the story. Communicating and fine-tuning the results, and taking decisions based on them, is still a different story. The complexity and comprehensiveness of probabilistic results require a new way of visualisation and interaction with the results. This has to be done in such a way that all stakeholders can be involved and easily understand the essence of the consequences of different alternatives for their own area of interest and that of the other stakeholders. The (further) development of serious games that allow stakeholders to visualise and discuss the effects of their own and each other's policy decisions (e.g., Savic et al., 2016, SIM4NEXUS, 2020) can offer a solution in this respect.

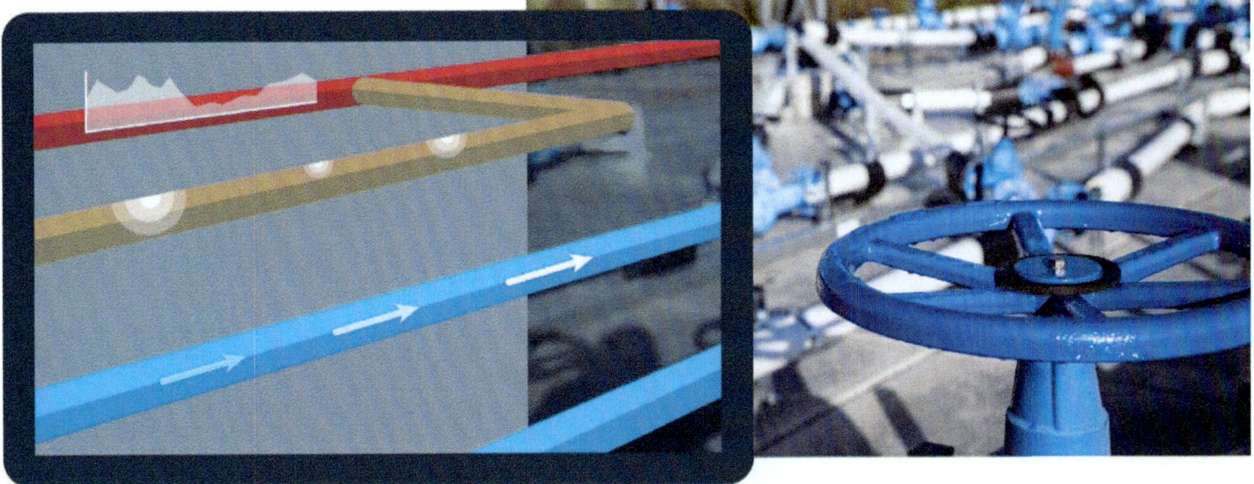

Figure 2.8 Augmented reality applications can project additional information on physical assets

Conclusions
Transforming data to information

This chapter has collected important basic concepts that are necessary to develop a successful Digital Water system. The number of tools available in the Digital Water toolbox is increasing every day, but a proper understanding of each stage of development and implementation is required to apply the right tools and obtain a useful, efficient, reliable system that fulfills the necessary tasks. Many Digital Transformation projects across the industry have failed simply because the data quality was not good enough to transform the data into useful information in order to reap the benefit of the insight that data can give to any water business. Quite simply, if we put "garbage in" then we will get "garbage out." This phrase has been attributed to many people but one of its first origins was by an army specialist, William Mellin, when working with computers. The phrase dates from the mid-1950s but is even more relevant today and, as the water industry moves towards Digital Transformation, its importance will only grow.

In this chapter we have discussed the principles of correct instrumentation and its necessary maintenance in order to obtain quality, reliable, and useful information. We have highlighted the importance of structuring our data and using the information about the data, the meta-data, to get the most out of the data and to extend its usefulness not only at the time of acquisition, but to be able to reuse it in the future when necessary. If we store not only the data but also information about how it was obtained or the specific operating conditions, in the future it may be useful for needs that have not yet been imagined. We have also introduced the fundamentals of uncertainty, i.e., the knowledge of the limitations of the data we collect. To what extent do we have to believe the data we measure or the results we get from a calculation or conclusion based on them? To what extent is the automatically drawn conclusion meaningful and relevant?

We have really only touched on these concepts and as an organisation becomes digitally transformed and becomes more digitally mature, the understanding of these fundamental concepts has to grow with the organisations as they become more and more digitally mature.

In the next chapter we will discuss the next steps in using the information that we have gathered with concepts such as Artificial Intelligence, Digital Twins and with resilience by using advanced control systems such as multi-variate process control. However, the danger is that if the data quality is not correct or the uncertainty of the data is not known, any operator is setting themselves up for, at best, failure and at worst operating by using systems that have a fundamental error associated with them. This inevitably will lead to poor performance and dangerous practices, even if very advanced processing techniques are used. If what a person learns has a fundamental error associated with it, then common sense can, eventually, correct that error. However, when we move into the world of a computer learning something fundamentally wrong, the error can be compounded, and the system as a whole is doomed to fail analogously to the error introduced into a calculation of the mean of a variable by the presence of an outlier among the sampled data.

Alternatively, if everything is based upon good quality data and with known uncertainty the benefits of Digital Transformation can be vast and can help the water industry to operate in vastly more efficient ways, saving energy, resources, and holistically helping the industry to achieve goals such as "net zero," and, therefore, helping the water and wider environment.

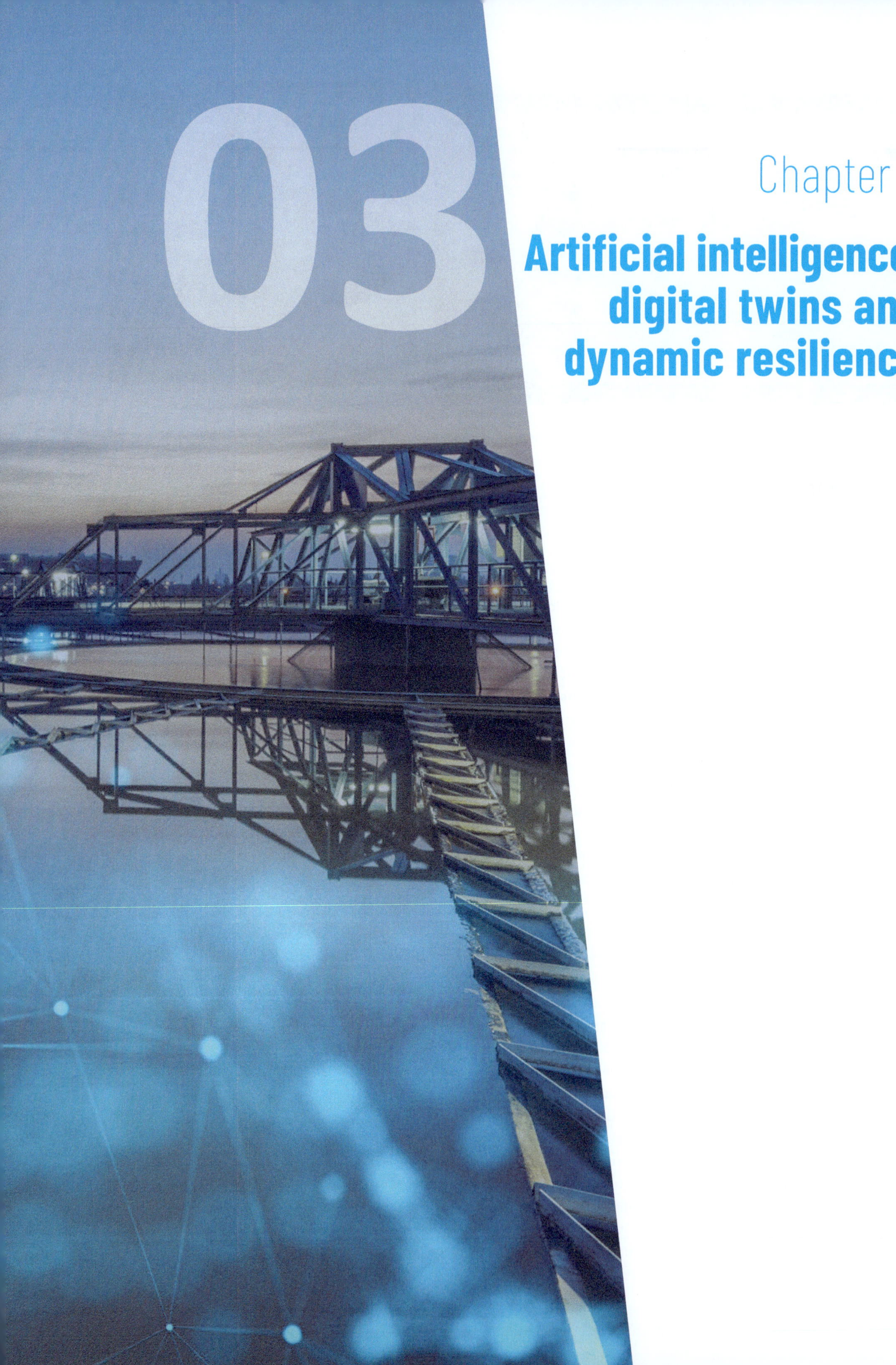

Chapter 3

Artificial intelligence, digital twins and dynamic resilience

Introduction

Global digitalisation and the accelerated development of digital technologies and analytical tools have ushered rapid changes in how humans interact with the different sectors of our society. This has had a significant impact on the international water industry, as described in this book. The amount of data we have at our fingertips is typically much more than any person or group can effectively use, so help is needed to improve our decision-making based on these large data sets. These vast datasets are generated intensively from instrumentation mounted in water/wastewater assets (see Chapter 2) leading to vast data silos of individually formatted and difficult to access data. For example, data logged by water companies is typically done at 15 min intervals, equating to 87,000 data points per annum. When multiple parameters are measured and logged for water and wastewater systems, it can result in millions of data points per annum. Processing this data can present a significant challenge for standard spreadsheet-based packages; therefore, new methods are required to convert this information into valuable insights.

The previous chapter discussed the importance of data and how to collect and transform it into information along with the challenges associated with these processes. When the data/information has been generated and stored, it should be used to generate insights that are both communicable and informed. These insights should extract otherwise unidentifiable characteristics, such as hidden dimensions or trends from pre-existing data or modelled outputs.

This chapter will present case studies where analytical tools (artificial intelligence, digital twins and dynamic resilience) were used to collect and better understand this available data. These analytical tools can help us understand data in distinct ways.

- Artificial intelligence (AI) helps us find and make use of obscured hidden patterns in large data sets, whether it be in numerical or visual patterns. This valuable tool enables computer systems to rapidly analyse water-related data in ways that human operators may not.

- Digital twins (DT) are essentially tools that force the integration of all the available data into a digital simulation of a system. This enables users to make more time-relevant and actionable decisions about operations by providing the bigger picture about how all the data streams are working together and are related to each other.

- Dynamic resilience (DR) uses machine learning to extract system operating conditions from actual WRRF data. Data-driven simulations are then performed before generated data is transformed into a dynamic resilience heat map of system stresses to represent how a water resource recovery facility (WRRF) system reacts to extreme events or stressors.

While the three tools are distinct from each other, they can be combined in some very powerful ways, as the case studies in this chapter show.

3.1 Artificial intelligence solutions for the water sector

Artificial intelligence (AI) is a term used to describe the theory and development of computer systems that are able to perform tasks normally requiring human intelligence. Examples of AI type systems include various expert systems, speech recognition, and systems used for financial trading. Two important concepts that are frequently mentioned in the context of AI are Machine Learning and Computer Vision.

Machine Learning is the study of computer methods and algorithms that improve automatically through experience. Machine Learning typically involves building some computer "black-box" model from data with the ability to make predictions without the need for additional software programming. One of the best-known machine learning methods is the Artificial Neural Network (ANN) that works by mimicking biological neural networks that exist in a human brain. ANNs learn from training data which is used to capture the functional relationships among the data, even if the underlying relationships are not known or the physical meaning is difficult to explain. This enables the ANNs to discover patterns in data that are often unknown, even to the best experts in the field. Computer Vision denotes a set of AI-type methods that are used to train computers to interpret and understand digital images and videos. Examples of computer vision applications include systems for facial recognition, medical diagnostics, and driverless cars.

Many AI methods exist and it is not our intention to describe these in detail. Instead, this section presents examples of AI-based solutions that make use of these methods.

AI-based solutions

REAL-TIME DETECTION OF PIPE BURSTS IN WATER DISTRIBUTION NETWORKS

Leakage is a major issue in water distribution systems worldwide. This example presents an AI-based system that detects pipe bursts/leaks but also equipment and other failures in these systems. The detection system works by automatically processing pressure and flow sensor signals in near real time to forecast the signal values in the near future (using ANN). These are then compared with incoming observations to collect different forms of evidence about the failure event taking place. The evidence collected this way is processed using Bayesian Networks to estimate the likelihood of the event occurrence and raise corresponding alarms (Romano et al., 2014). The system effectively learns from historical burst and other events to predict future ones. Elements of the detection system, developed initially as part of a research project, were built into a commercial Event Detection System (EDS).

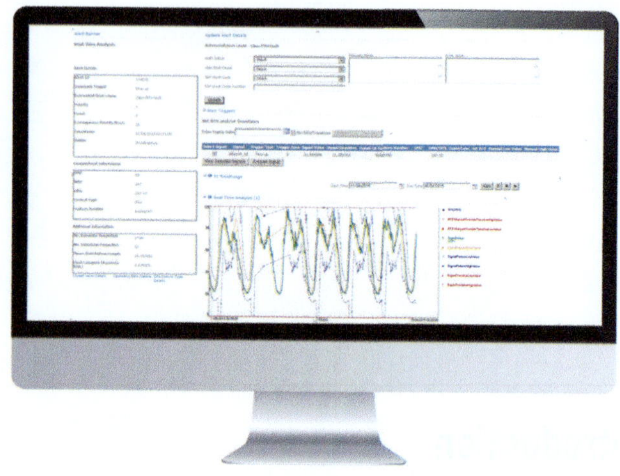

Figure 3.1 EDS Screenshot with Real-Time Analysis of a Specific Alarm

An EDS has been in use by a large UK water company since 2015. It processes data from over 7,000 pressure and flow sensors every 15 minutes (see Figure 3.1). This enables EDS to detect pipe bursts and related leaks in a timely and reliable manner, i.e., shortly after their occurrence and with high true and low false alarm rates. In addition to detection, EDS can proactively prevent burst events by detecting equipment failures that often precede these events (e.g., pressure reducing valve failures). An EDS does not make use of a hydraulic or any other simulation/mechanistic model of the analysed water distribution network, i.e., it works solely by extracting useful information from sensor signals where bursts and other events leave their imprints (i.e., deviations from normal pressure and flows signals). This makes the EDS robust and scalable as it enables data to be processed in near real time (i.e., within the 15 minute time window). The use of EDS has resulted in major operational cost savings to date, significantly reduced customer supply minutes lost, reduced leakage and several other benefits to the water company (full details not mentioned for commercial reasons). All this has led to a change in company business culture and improved service to over 7 million customers.

AUTOMATED ASSET CONDITION ASSESSMENT USING AI AND COMPUTER VISION

The inspection of urban drainage (i.e., sewer) systems' pipes is important as undetected structural and other faults (e.g., displaced joints, cracks, etc.) may result in severe pollution and/or flooding incidents. This inspection is done usually by recording CCTV videos and then analysing these manually. This process is time consuming (i.e., costly) and subjective/inconsistent in nature hence not necessarily always reliable.

The AI-based solution automates the process of analysing CCTV videos and detection of faults in pipes. It does this by using computer vision and machine learning methods (Myrans et al., 2018). Image processing is conducted first to process and convert the CCTV images into suitable data. This data is used then to detect faults with a help of a Random Forest machine learning method. This method is trained before it is used on a number of pre-labelled CCTV images. The automated detection works similarly to the human face recognition system although the task of fault detection is more complex in sewers due too many different types of faults that exist and that can manifest themselves in very different ways in CCTV images.

The above solution was successfully evaluated and validated on unseen real CCTV data from several water companies in the UK, Finland, and Australia. It has a high true detection rate accompanied with low false alarms rate. This technology is currently being commercialised by a variety of companies around the world.

PREDICTIVE WASTEWATER TREATMENT PLANT CONTROL

Royal Haskoning DHV's Aquasuite® software was deployed at PUB Singapore's Integrated Validation Plant at Ulu Pandan Water Reclamation Plant in March 2019. It was introduced to provide operators and managers with predictive insights while improving plant performance. Real-time data is collected on the plant's flows and qualitative measurements, including those for ammonia, nitrates, oxygen, phosphates, and dry solids, and builds a historical database. The software then makes use of advanced analytics and Machine Learning algorithms to predict the plant's wastewater flows and loads, oxygen needs, chemical dosing needs, and other requirements (see Figure 3.2). The system controls key treatment processes, automatically optimising them in real-time based on its predictions and the plant's historical performance.

It further detects anomalies in the plant's processes through quantile regression techniques based on multiple measurements. Prediction accuracy of the influent flow increases over time as the software is learning, reaching a prediction accuracy of 88% after just one month.

Connecting to the plant's Supervisory Control and Data Acquisition (SCADA) system to gather data and control key processes, data is sent to the cloud component of this solution. With the data collected in near real time in the cloud, the software tracks the actual performance of the on-premises solution through a digital twin, a digital replica of a physical system. More advanced analytics are made available to operators through the cloud, while the on-premises part keeps optimising real-time performance. Machine learning is used to understand the efficiency of each process, and the learnt relationship is then used with the prediction of the influent load to decide the most optimal efficient set points for treatment. Results show that this digital twin learns and predicts operations several days ahead and it can function as an autopilot, able

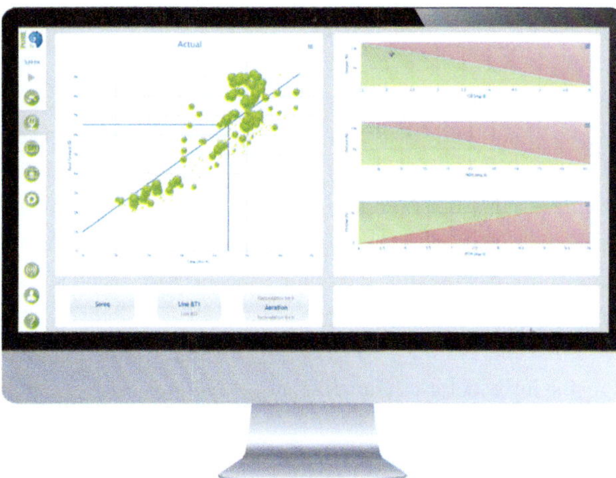

Figure 3.2 Dashboard view of Aquasuite Pure

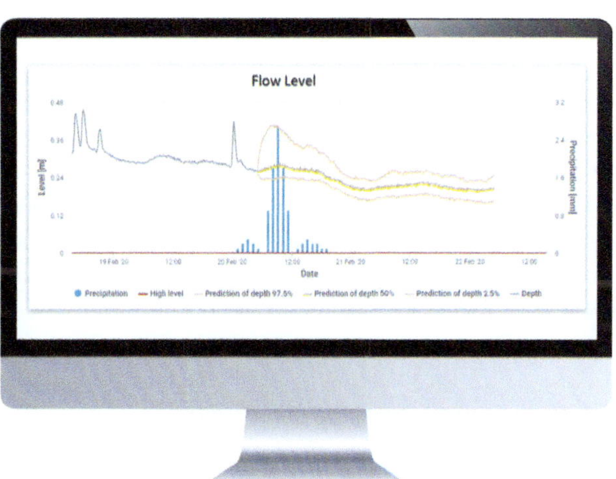

Figure 3.3 Dashboard view of flow levels and predicted outcomes

to perform unattended operation. Preliminary results show a reduced aeration flow of up to 15% with predictive control, resulting in corresponding energy savings.

SMART ALARMS FOR PROACTIVE WASTEWATER NETWORK MANAGMENT

The flow prediction component of this digital twin is an AI-powered predictive analytics tool for wastewater networks that enables prediction and early warning of critically high-water levels in sewers, potential pollution events, and detection of anomalous levels that could indicate blockages in the network (see Figure 3.3).

Water utilities are most at risk for blockages during rainfall events, in part because of sewer flooding, capacity limitations, and foreign objects.

AI is used to detect when high level flows in sewers exceed critical levels or are not consistent with the expected flows. The system identifies these as anomalies through a customisable smart alarm notification system. Using AI in this context drastically reduces false alarms and prioritises the remaining alarms to manage blockages accurately and efficiently through early detection and real-time response, preventing further problems like fatbergs. Monitors located strategically around the catchment collect flow and level data. That data is transmitted to a secure, reliable cloud platform that displays the information in easy-to-use dashboards and reports for analysis. Using smart algorithms, real-time data, historical data sets, and information from other sources such as rainfall, data are translated into actionable insights for operators.

The flow prediction AI model is built on two multi-layer perceptron ANNs that work together to support operators. A regressive ANN predicts flow levels for the next 48 hours and a heuristic ANN identifies event anomalies like blockages. Integrating machine learning allows the anomaly detection to continually improve over time. Human data technicians review the data alongside the predictive analytics software to verify the alarm protocols. The additional human intervention helps detect any existing or temporary anomalies, seasonal variations, and external influences. Prediction of a high sewer level enables water utilities to proactively inform their customers of potential sewer flooding, and, in critical environmental regions, to actively prevent spills. Anomaly detection enables the utilities to detect and clear blockages before they cause sewer overflows.

REAL-TIME FORECASTING OF SEA CURRENTS

Over the past few decades ANNs have evolved in a popular approximation and forecasting tool, frequently used in a range of problems and application areas (ASCE Task Force, 2000). In this solution, a recurrent ANN is used to create a real-time hybrid data assimilation system resulting in extremely accurate forecasts of sea surface currents. This solution was developed to support the construction of Øresund Link connecting Danish capital Copenhagen and City of Malmø in Sweden (Babovic et al., 2001). The combined roadway and rail line bridge run nearly 8 km where it then transitions into an underwater tunnel for the remaining 3.5 km. Due to the material of the seafloor, a tunnel was not possible. Instead, engineers chose to sink and connect 20 prefabricated reinforced concrete segments – the largest in the world at 55,000 tonnes each – and interconnect then in a trench dug at the seabed. The elements were prefabricated in a special purpose-built facility north of Copenhagen, sealed shut and using a specially designed barge along with seven tugboats, lowered into place at required accuracy of alignment of 2.5 cm. The towing operation for each element could be conducted within a "window of opportunity" of 36 hours during which sea surface currents had to be guaranteed to be less than 0.75 m/s. Despite extremely challenging conditions, all 20 elements of the Øresund link's tunnel were successfully placed at their positions

between 11 August 1997 and 6 January 1999 – 20 towing operations in 17 months! It is suggested that the accurate ANN-based forecasting of seal surface currents was one of the key factors in this achievement. The Øresund Strait Link was opened to the public on 1 July 2000.

BAYESIAN NETWORKS FOR PRO-ACTIVE ASSET MANAGEMENT

The economic and social costs associated with pipe bursts and leakage in modern water supply systems are rising to unacceptably high levels. The challenge for the decision-maker is to determine what pipes in the network to rehabilitate, by which rehabilitation method, and at what time within the planning horizon. Advanced AI, machine learning, and statistical methods are used in order to establish risks of pipe bursts (Economou et al., 2014). For example, analysis of the database of already occurred burst events can be used to train a risk model as a function of associated characteristics of bursting pipe (its age, diameter, material of which it is built, etc.), soil type in which a pipe is laid, climatological factors (such as temperature), traffic loading, etc. In this context, a machine learning model based on a Bayesian network is used to predict which pipes are most vulnerable to failure including a metric for failure probability (Babovic et al., 2002). Bayesian networks are a probabilistic graphical model that use Bayesian inference for probability computations. The approach models conditional dependence and, therefore, causation. Through these relationships, one can efficiently conduct inference on the variables in the graph exemplifying pipe failure mechanism. Pilot projects using the approach have been conducted in Sweden, Singapore, UK, and Denmark.

COMPUTER VISION FOR OPPORTUNISTIC RAINFALL MONITORING

The quantity and quality of precipitation data are crucial in meteorological and water resource management applications. Rain gauges are a classic approach to measuring rainfall. However, as we enter the age of the Internet of Things (IoT) in which "anything may become data" so-called opportunistic sensing using unconventional data sources offer a promise to enhance the spatiotemporal representation of existing observation networks. One particular area attracting attention is the estimation of quantitative and analytical rainfall intensity from video feeds acquired by smart phones or CCTV surveillance cameras. Technological advances in image processing and computer vision enable extraction of diverse features, including identification of rain streaks enabling estimation of the instantaneous rainfall intensity (Allamano et al., 2015). Recent AI and machine learning approaches rely on the use of autoencoders, deep learning, and convolutional neural networks to address the problems. Companies such as WaterView (Italy), Hydroinformatics Institute (Singapore), as well as universities (e.g., Southern University of Science and Technology China, Shenzhen) have proposed and implemented practical approaches to weather hazards in energy, automotive and smart cities application domains (Jiang et al., 2019).

3.2 Operational digital twins in the urban water sector

Due to advances in instrumentation and the increasing availability of online data and computing capacity for utilities (e.g., via cloud computing), the development of digital twins has recently attracted large interest in the urban water sector. This interest is primarily driven by two trends in our industry:

1) The increasing amount of data available at our facilities; and

2) The business drivers around both improved capital and operational efficiency targets.

The amount of data available at many facilities now exceed the amount that most operational staff can utilise in their day-to-day operations. Digital twins can take those data and present them in terms that staff can use effectively on a day-to-day basis (Garrido-Baserba et al., 2020). The second driver towards efficiency relates to the ability of a digital twin to go beyond conventional PID (i.e., proportional–integral–derivative) control, using model-based control and data-driven optimisation. This improvement has implications for capital expenditure since more reliable control can result in reductions in conservatism, and thus lower capital expenditure (Stentoft, 2020).

In a broad sense, "digital twin" refers to a digital replica of physical assets. Within the water sector, digital twins are combinations of models that provide a digital representation of a specific part of the water system (e.g., water resource recovery facilities, sewers, etc.) and utilise automated real-time data from multiple sources to, e.g., simulate expected, desired, or critical behaviour of the physical system (Pedersen et al., 2021).

While there is agreement in a broad sense about the definition, there is a lack of consensus on which elements characterise a digital twin. Some experts consider traditional mechanistic models with frequent recalibration using sensor data enough to qualify as a digital twin, whereas others consider the interaction with real-time data as a key element in an operational digital

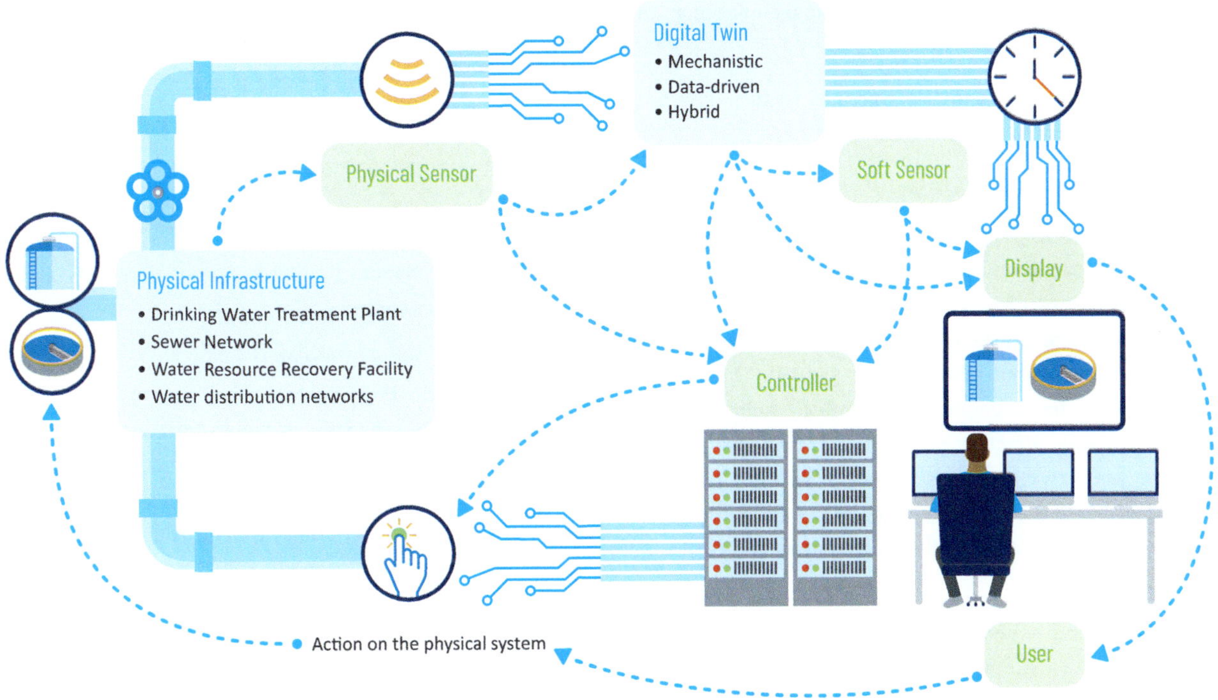

Figure 3.4 Basic structure of a digital twin application. Elements in blue are basic elements needed for implementation of a digital twin, while elements in green are complementary to the digital twin, ultimately to enable an action on the physical system.

twin. In this discussion, it is important to distinguish between digital twins for operational use (relying on near- real-time data from the physical system) and digital twins for planning, design, construction, or investment purposes (relying on historical data from similar physical systems or expected demands given the operational environment of the physical system under development). In some cases, digital twins are considered as soft or virtual sensors of the physical system. Thus, aspects like model complexity (mechanistic, data-driven, or a hybrid combination of both), handling of uncertainty and data requirements, among others, need proper assessment to ensure successful alignment between digital twins and their intended application. Furthermore, data assimilation from different data sources remains a key challenge to be addressed. Figure 3.4 illustrates the most basic structure of a digital twin and its interaction with the real system and users.

Digital twin applications include (but are not limited to):

(1) Investment planning via future scenario analysis, traditionally performed with models calibrated with historical data;

(2) Real-time decision support for selection of different operational strategies;

(3) Operator training;

(4) Online optimisation (e.g. model predictive control) for energy or resource savings or compliance management (e.g., to minimise carbon footprint);

(5) Asset management and interaction with different stakeholders; and

(6) Sewage epidemiology (Poch et al., 2020).

Within the water sector, digital twins can support the transition towards more proactive management of water infrastructure, whereby different processes and systems can be operated to mitigate disturbances before they have adverse impacts on performance (Karmous-Edwards et al., 2019). Thus, there is large potential for economic savings (e.g., online energy optimisation), more effective protection of the environment (e.g., model predictive control for effective nutrient removal) and increased benefits for society (e.g., improved storm water management to minimise risk of flooding in urban areas).

In this section, two examples of online digital twins for operational decision support are presented: one focuses on improved control of combined sewer overflows (CSOs) in a sewer system and the second focuses on proactive maintenance and process optimisation of a WRRF intended for water reclamation. Both are considered hybrid digital twins which uses a combination of mechanistic modelling and machine learning computational methods.

Digital twin for sewer networks

This case study describes the digital twin approach used for sewer networks in the Swedish cities of Gothenburg and Helsingborg within the project Future City Flow. The approach is similar in both cities, but this case study will focus on the digital twin developed in Gothenburg (WRRF: 900 000 PE; catchment: 240 km² with 20 km² of impervious surface; 40% combined sewers).

The utilities regularly experience high flows in the sewer collection systems leading to spills from CSOs that lead to 3 billion litres per year (2.2% of total flow) of untreated wastewater being discharged to the environment. The CSOs mostly occur due to large flow variations caused by heavy rainfall events. To manage issues related to CSOs and to reduce storm weather impacts on the WRRF, an operational digital twin approach was envisioned as a decision support system with online flow prediction and suggestions for control strategies (the final decision is currently made by the operator). Part of the last stage of the project (2019- 2021) focuses on implementation of full model predictive control (MPC).

COMPONENTS

The structure and different components of the digital twin are shown in Figure 3.5. The main part of the digital twin consists of a model of the sewer system built in MIKE URBAN using several different modules, including:

(1) A dynamic model for the hydraulics of the pipes and tunnels;

(2) Conceptual hydraulic models to describe sub-catchments;

(3) Optimisation modules for real-time control; and

(4) Modules for handling of precipitation forecasts and rain gauge data quality control.

The model has been calibrated manually for many of the sub-catchments in the network where rain gauges and flow measurements are available, as well as for the hydraulic model of the main tunnel system. The catchment model is re-calibrated manually to achieve better flow prediction as more sensors are added to the system over time, thus providing greater spatial resolution in the system for model calibration.

The physical infrastructure connected to the digital twin consists mainly of flow and water level sensors in the central part of the sewer system and at the WRRF that provide online, real-time updates of these measurements, as well as status from actuators (pumps, valves, and gates) in the system.

The operator visualises simulation results on a dedicated website, separate from the SCADA system, showing flows in

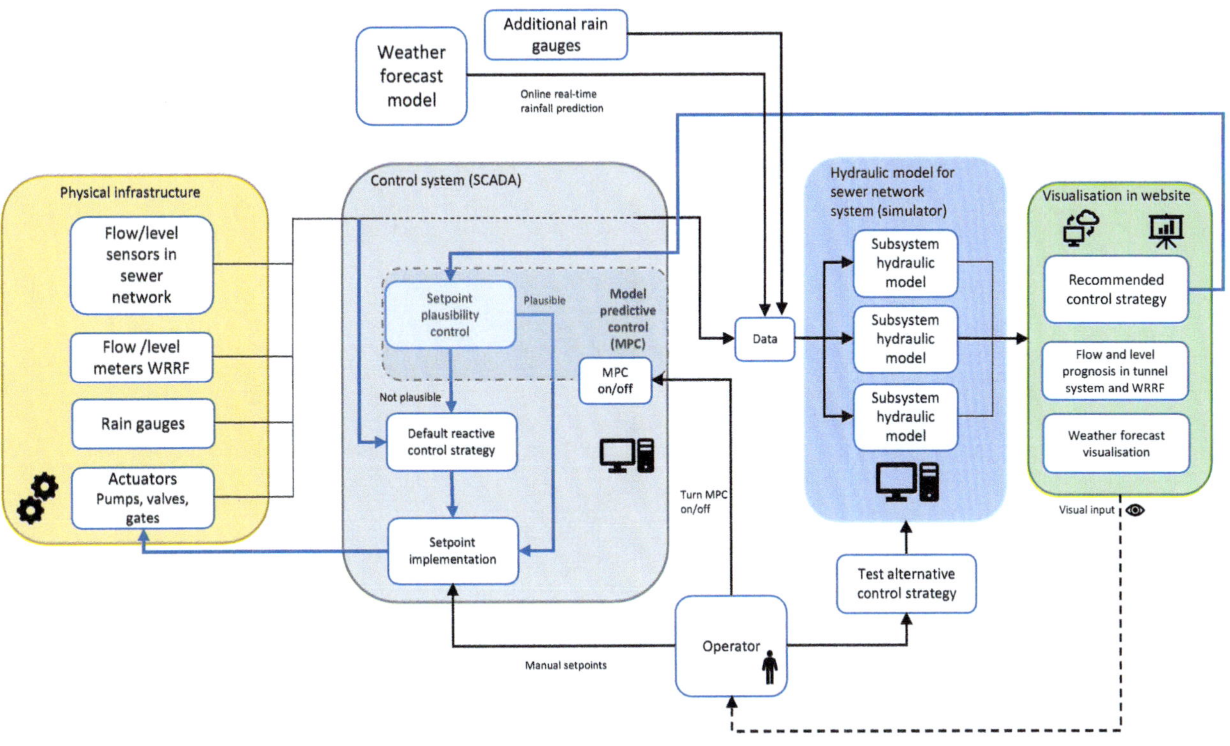

Figure 3.5 Sewer network digital twin structure and components in the current implementation as well as with model predictive control (work is now underway with the goal of implementing full real-time model predictive control).

the catchment, influent flow to the WRRF, and a comparison of predicted flows with both the default reactive control strategy (based on an empirical level-flow relationship) and a recommended control strategy (provided by the simulator) (Figure 3.6).

Both sets of predictions are for the next 2.5 days, based on the latest weather forecast. The recommended control strategy is updated every hour, resulting in new recommended set points. The operator then has the option to implement the recommended control strategy or to evaluate their own strategy in the simulator and evaluate those results. The future MPC will include plausibility control of the recommended control strategy before final implementation. If the plausibility check fails, the controller will revert to the default reactive control strategy.

To determine the recommended control strategy, the controller uses an algorithm to optimise a pumping scheme to achieve inflow to the WRRF that is as constant as possible during the next 12 hours, while considering boundary conditions, such as allowable water levels in the sewer (to avoid CSOs), pumping capacity, pumping conditions etc. To avoid discrepancies between current measured values of flows and water levels in the sewer network and pumping stations and the corresponding values in the model, corrections are required. Data assimilation techniques are therefore used to initialise the model using real-

time data from the sewer network before each simulation and to adjust the forecast according to identified patterns of deviations between measured and simulated values.

CHALLENGES

Predicting the future in such a varying system is always a challenge. Since accurate flow and level predictions depend on accurate rain forecasting, this has presented a major challenge due to the uncertainty of weather forecasting models.

Another challenge is to represent variations in extraneous water from different sources (such as direct inflow, infiltration water, drainage water, etc.) for different hydrological events both in the short term and the longer term. This requires a modelling approach that allows an accurate representation of the urban hydrology and of extraneous water impacts on the sewer system, not only for single events but continuously in time, using both hindcast and forecast data. Figure 3.6 illustrates the effect of the prediction time horizon on predicted flows by showing past predictions of WRRF inflow with two prediction horizons (0—1 h and 3—6 h in advance). The curves display the mean predicted flow values for the specified interval after the time of forecast which are saved at the time in the middle of the interval (i.e., the curve tagged "prediction horizon 0—1 h" shows the prediction made 0.5 h earlier for the mean flow during the next hour after

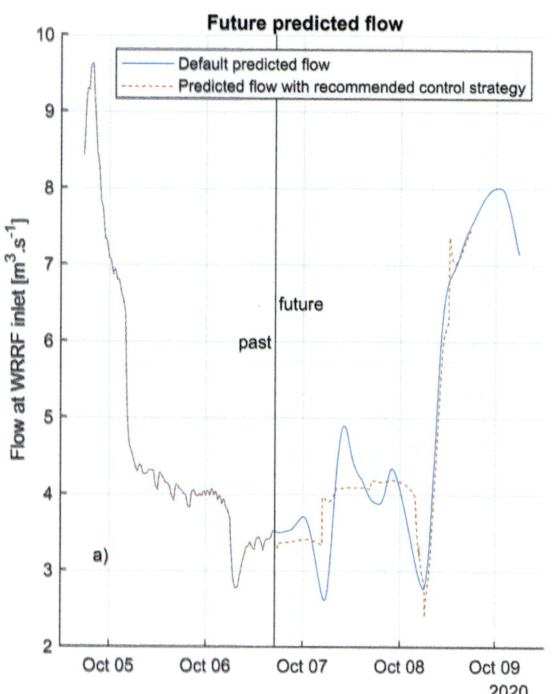

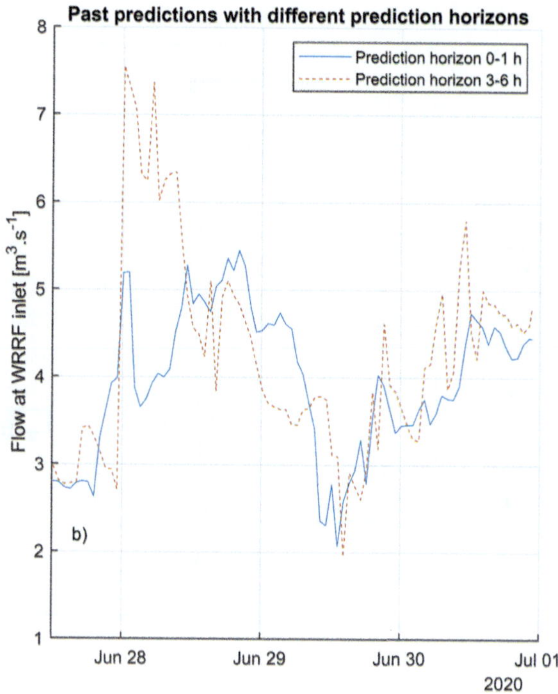

Figure 3.6 (a) Example of model predictions for influent flow to the WRRF using the default control strategy (solid line) and the recommended control strategy suggested by the model (dashed line); (b) Example of past model predictions with different prediction horizons (0 — 1 hours and 3 — 6 hours).

the time of forecast, while the curve tagged "prediction horizon 3—6 h" shows the prediction made 4.5 h earlier for the mean flow in the period 3—6 h after the time of forecast).

A large rainfall event was initially predicted to occur on 28 June, as suggested in the 3—6 h curve, but the rainfall volume and WRRF inflow were predicted to be much smaller when predicting 0—1 h in advance.

Data management has also been a challenge, mainly to create a secure and robust data transfer. Specific issues include time stamps when combining data from different systems (e.g., because of daylight savings time); ensuring data quality for precipitation data; managing the size of files for weather data which was initially too large for efficient transfer; and integration of the SCADA system, which normally only handles historical data, with local databases and interfaces for storage and visualisation of future predicted scenarios, respectively.

A critical part of the digital twin is the connection between the user and the models. An important aspect has been to develop confidence in the digital twin among the operators in the control room. A model that is sufficiently accurate and whose limitations are well understood is therefore important. A related issue has been the question of how to display uncertainty in predictions to the user (since weather predictions can include many different scenarios and many pumping strategies can be tested).

RESULTS AND BENEFITS

The digital twin allows for increased confidence in decision-making as the effects of a change in control strategy can be visualised quickly. Simulations of the real-time control strategies indicate that CSO spill events can (ideally) be reduced by 50% in the Gothenburg case. In Helsingborg, CSO spill events have already been reduced by 32% before implementation of MPC, by thoroughly studying the sewer system and exploiting opportunities that emerged. Benefits at the WRRF include:

(1) A more constant influent load which leads to more stable treatment processes;

(2) Lower risk of reaching critical load situations (e.g., by pumping water from the tunnel system before large flows are expected, so that more volume is available for attenuation, thus avoiding large flow peaks that can be hard to handle); and

(3) An increased margin for handling issues with pumps or accumulation of material in the coarse screens at the inlet of the WRRF. A major benefit with MPC is that the control system takes into account both current conditions and predicted future conditions.

Digital twin for a water resource recovery facility

Jacobs partnered with PUB, Singapore's National Water Agency in an R&D project to create a digital twin of the Changi Water Reclamation Plant (CWRP). This digital twin provides new insights into the ongoing operations and maintenance of the facility, supporting increased productivity and enhancing operational resilience.

The digital twin is envisioned as an advisory tool without direct control capabilities, which can grow into control functions as staff gain confidence in the tool. It has automated data inputs directly from both the SCADA system and the laboratory information management system (LIMS), as well as auto-calibration and soft sensor capabilities. This model is expected to assist PUB in simulated scenarios to test and calibrate strategies to enhance the plant's water quality as well as optimise its energy and chemical consumption. The use of the model is also in line with PUB's goal of tapping smart technologies to increase productivity and improve resilience in operations. A process flow diagram of both CWRP and the digital twin simulation is shown in Figure 3.7.

COMPONENTS

The digital twin includes dynamic semi-mechanistic models of the CWRP whole plant hydraulics, controls, and processes. Hydraulics and controls are simulated within Jacobs' Replica™ simulator. Processes are modelled using Dynamita's Sumo© whole plant simulator with direct communication between Replica and Sumo. The inputs to the digital twin are only those currently available at the facility. Data-driven (machine learning) influent predictions are used as part of the digital twin functionality for predicting plant performance up to five days into the future.

The influent soft sensor workflow of the digital twin is illustrated in Figure 3.8. Input data to the model are first conditioned by removal of "bad" data, defined as negative or zero data that are not consistent with other related data. A historical test is then made to remove all data that are not within normal variations ($\pm$ 2.5*log normal interquartile range, IQR). The digital twin inputs from SCADA are the various flows measured in the facility, the online primary effluent ammonia values, air rates to the various bioreactor zones, and other relevant operational set points. From the data, in combination with the LIMS data, a dynamic raw sewage influent file is generated, thus creating a soft sensor of the actual influent to the facility. Air rates and operations set points read from SCADA are directly input into the digital twin.

The initial calibration of the digital twin was done manually; control and hydraulics calibrations were based on actual measurements. The process calibration was first done in a steady state, then dynamic calibration was done based upon the first 6 months of a full data set. The digital twin also included limited auto-calibration while running. This was accomplished as measurements became available where, for example, primary effluent laboratory total suspended solids (TSS) and chemical oxygen demand (COD) data were used to calibrate both the primary clarifier TSS removal and the soluble COD/COD fraction in the raw sewage based on the most recent performance data.

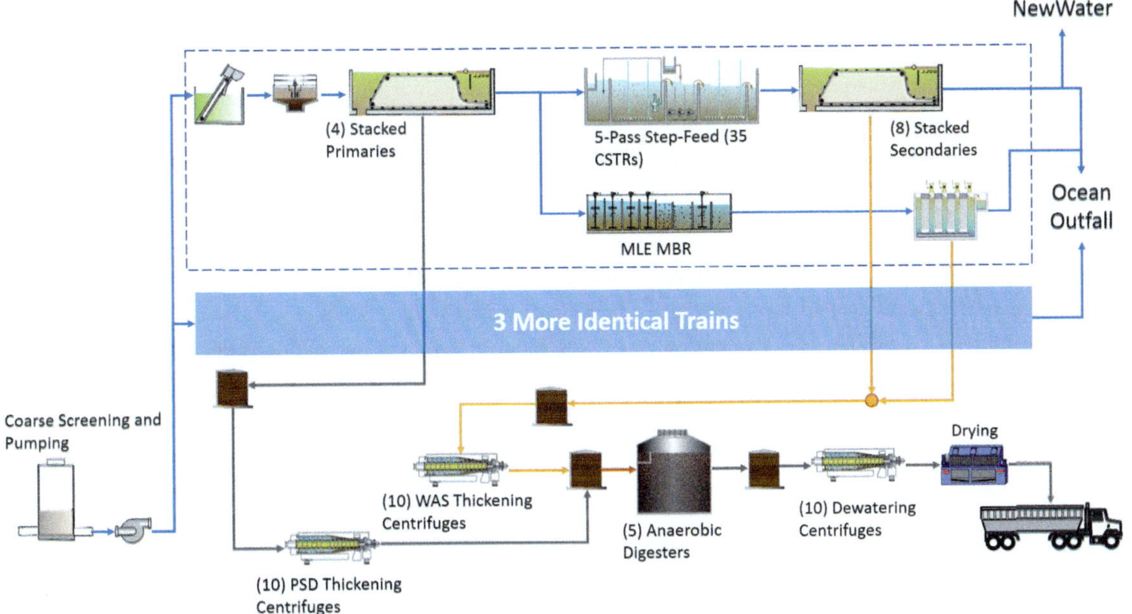

Figure 3.7 Process flow diagram of Changi WRP and scope of digital twin simulation (inside dashed line). The digital twin includes the four water treatment trains.

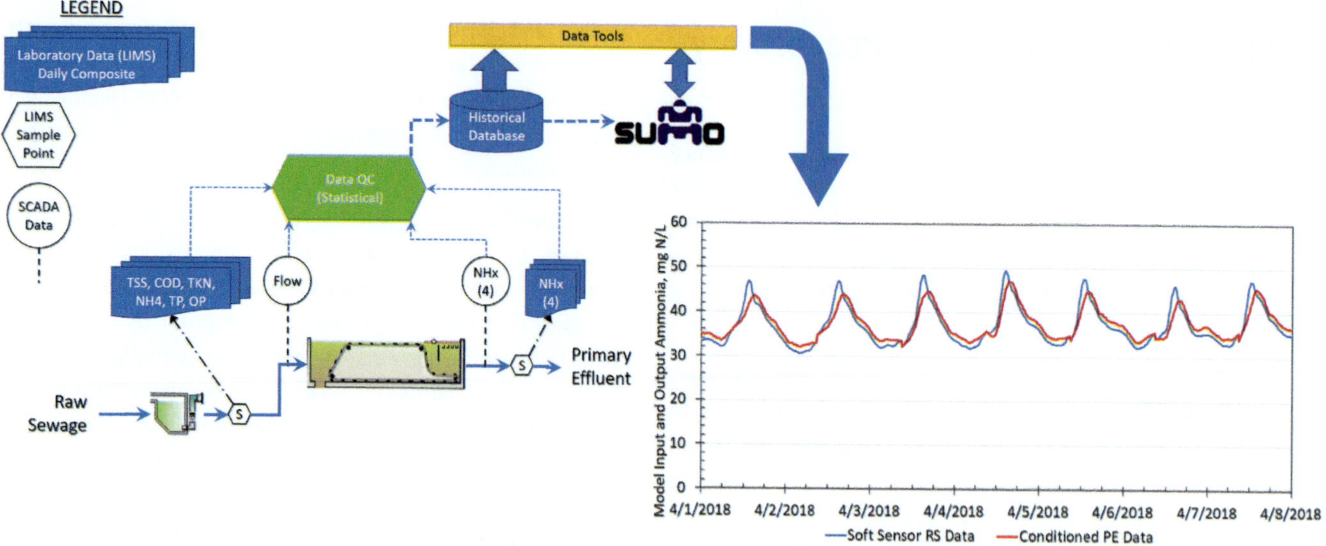

Figure 3.8 Data flow schematic within digital twin for influent load development for the influent "soft sensor"

CHALLENGES

The largest challenge in this project is speed of computation, as results are needed for both current operation/model comparison functions, scenario evaluation, and future predictions using Monte Carlo techniques. This must be accomplished within 24 hours of receiving the relevant data. For example, the process (Sumo) model is over 40 MB in size and the Replica hydraulics and control model is over 170MB in size, thus giving an indication of the model complexity.

RESULTS AND BENEFITS

There are three primary functionalities of the digital twin. First, comparison between model predictions and measured data can be used to highlight areas which require particular attention by operators and maintenance staff, thus minimizing their efforts. For example, a warning will be issued if one of the primary effluent online ammonia probes differ significantly from the others and the model or if the laboratory primary sludge total solids does not match up with the model results.

For the latter, a check of the primary sludge flow meter or confirmation of the laboratory results would be required. The second use of the calibrated model includes evaluation of various operational scenarios, both operator-defined, and fixed. Lastly, the auto-calibrated model is being used to predict the likelihood of future events at the wastewater facility up to five days in the future, a "wastewater weather forecast" that operations can use to help proactively operate the facility.

Conclusions

The case studies presented in this work demonstrate the water industry's eagerness to move to digital solutions such as digital twins to improve the performance of infrastructure. This section demonstrates that large potential savings can be derived from better process automation, online optimisation, fault detection, maintenance, and more proactive operation. However, there is no "one recipe fits all" for digital twin applications, and this may hinder the rate at which the full potential of this tool can be realised. Good practice protocols need to be established to support digital twin developers. The protocols should provide answers to the following questions:

- **Purpose of the digital twin:** what are the objectives and the expected gains? Can the objectives be attained with the existing infrastructure and available data?

- **Data:** what is the impact of data quality and quantity on the applicability of a digital twin? How to feed data to it? How to calibrate it?

- **Model:** how to identify the right modelling approach for a given objective? What is the right level of detail (granularity) in a model? How to be confident that model predictions are acceptable for a digital twin? How to manage model uncertainty?

- **Robustness of the digital twin:** how does the digital twin continue to evolve? How to perform model validation using available online data?

3.3 Digital dynamic resilience for wastewater treatment processes:
Exploiting real data for long-term resilience

With societal (e.g., COVID-19) and climatic factors coinciding with increasing regulatory pressures, the resilience of wastewater networks and infrastructure is reducing globally. Historically, water companies have relied in reserve capacity, but are now being forced to manage extreme dynamic responses as wastewater assets react new stressors. An example of this is rainfall intensity, which has already increased from 12 to 24 % (Fischer et al., 2014) and has been commonly followed by prolonged dry periods driven by climate change (NASA, 2016). These dramatic variations generate complex dynamic changes and can drastically reduce the resilience of networks and, in turn, of the network water resource recovery facilities (WRRF). Without unified quantification methods, it is impossible to compare the resilience of different wastewater processes or systems, when exposed to climate and societal stressors. Also, the complexity of dynamic changes makes it virtually impossible to simulate the numerous dynamic factors that combined cause these reductions in resilience.

To avoid possibly cumbersome modelling and simulation of possible scenarios, dynamic resilience uses actual WRRF data to visualise zones of process stress and resilience as a heat map. The methods presented in this section separate stressors present in water company data as the 'cause' of an event, and the 'effect', whether a WRRF experiences process stress or resilience as a result. This separation of stressors and process stress is key to isolating the cause of an event then its manifestation as the effect. This separation of the stressor and process stress requires data feedback from WRRF process and systems. Data generated by water companies is ideal for computation of dynamic resilience in response to extreme events, where the cause (stressor) and effect (process stress) can be seperated. This data is generated in vast quantities daily (typically < 1 h intervals) and is used to make operational decisions, but often remains in silos, or as described by Aguado et al. (2021), 'data graveyards'. Another challenge is that WRRF data can be difficult to interpret when instruments are poorly maintained or installed incorrectly (Grievson, 2020), but meaningful observations can still be made to interpret resilience metrics. Therefore, this data must be exploited to understand the dynamic resilience. Without data (real-time or from silos) connecting a WRRF or process to resilience, evaluations may remain theoretical and iterative, which is computationally intensive.

This section starts by describing the historical context of resilience, before moving onto the dynamic resilience of wastewater processes and WRRF. Real-world examples of societal and process related resilience from industrial and academic experts are provided, which discuss the challenge of generating data under uncertainty of ageing infrastructure. A case study is then presented on dynamic resilience using actual WRRF data. The case study shows how actual water company instrument data could be used to evaluate stressors and process stresses independently. The outcomes of dynamic resilience case studies are then presented as a series of contoured heat maps.

This section covers the fundamental and historical context of resilience and its evolution within the water industry globally

THE FUNDAMENTALS OF RESILIENCE FOR WASTEWATER TREATMENT

The word resilience dates back to the 1620s, described as the 'act of rebounding', i.e., the ability to recover from an external event (Harper, 2019). By definition, the term rebound indicates that an external event causes a system to deviate from its initial reference position, and must be reconciled for an event to be considered complete. Classical resilience theory, originated from social sciences, focusing on ecological resilience in predator-prey systems, where a system absorbs change until moving to a completely new state as described by Canadian ecologist Crawford Holling (Holling, 1973). This description was later expanded to include engineering resilience, which focuses on maintaining stability close to an equilibrium stready state. An example of engineering resilience is a set point, where the system is controlled to maintain predictable operation close to user defined optimum.

Wastewater systems can be a combination of the two types of resilience (ecological and engineered). Extraneous events shift the wastewater collection systems to a new operating state and, as result, the system performance is controlled, in order to manage undesirable process upsets or disturbances. Using the example of an activated sludge (AS) system, the microbial ecology is controlled by an engineered system of mechanical/electrical components (aeration and pumps). Therefore, a more accurate classification of a biological treatment process would be engineered ecology, where process engineers apply engineering principals to control biological ecology. Wastewater catchments can also behave ecologically, where systems can move to a completely different operational state as populations vary, or extreme events occur. Many of these changes in the catchment have a direct effect on the WRRF engineered ecology due to dramatic operational state changes which reduce resilience by

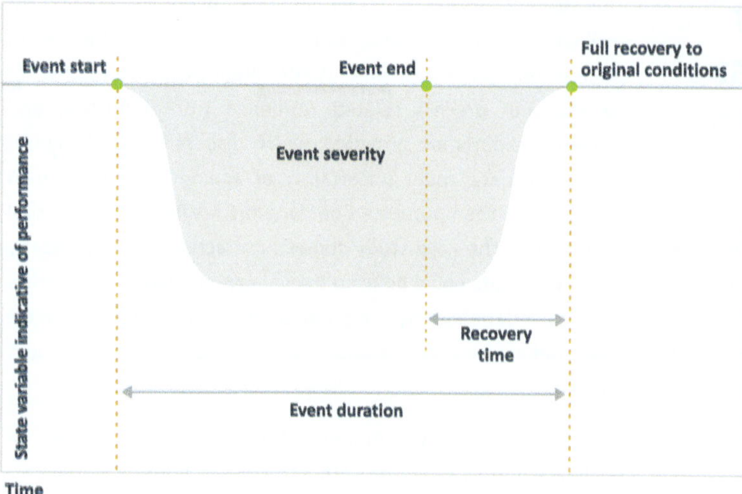

Figure 3.9 Resilience to a stressor presented by Juan-García et al. (2017) adapted from Mugume et al. (2015)

eroding operational safety factors (reserve capacity). Therefore, to evaluate the dynamics of resilience, we must understand that we are actually managing engineered ecology (i.e., engineered systems that aims to control microbial ecology). This is where, dynamic resilience extends existing theories: it accepts that dynamic changes in resilience are both, positive (resilience) and negative (stress).

Another difficulty for resilience has been selecting an appropriate definition. A commonly accepted definition was proposed by Walker et al. (2004) as:

'Resilience is the capacity of a system to absorb disturbance and reorganise while undergoing change so as to retain essentially the same pre-disturbance process, form, identity, and feedbacks'

This definition suggests that a system undergoing change adapts to an event before returning to its original condition (Figure 3.9). However, the definition does not account for the complexity associated with biological wastewater systems, which have numerous states (ecological and engineered), some temporary and some permanent. Resilience is also reduced further, by additional novel extreme states occurring outside of diurnal and seasonal patterns, which related to climate change and modifications to human behaviour. When normal operation is combined with novel stressors, simulation of their response is cumbersome, vast numbers of theoretical iteratons. Therefore, if WRRF data has sufficient resolution (adequate range and quantity), it could hold clues to these novel states, where many parameters are affected by stressors, and the process stress/resilience response (dynamic resilience).

There are many interrelated factors must be considered to evaluate the dynamics of resilience. Therefore, it is crucial to not only consider wastewater infrastructure and WRRF, but also factors that lead to stressors under the following headings:

1. **Political resilience**
2. **Economic resilience**
3. **Social resilience**
4. **Technological resilience**
5. **Environmental resilience**
6. **Legal resilience**

PESTEL factors and their interactions are responsible for many of the stressors exerted on ageing assets and infrastructure. However, in times of societal adaptaion and change, numerous PESTEL factors can become interrelated, generating significant complexity and uncertainty. Nevertheless, pre-existing clues of dynamic stressors and process stresses are apparent in water company transactional data and surveillance data used by government agencies to monitor COVID-19. This section focuses on stressors occuring from a wastewater catchment that have a significant effect on the stresses generated within a specific WRRF. As shown in Figure 3.10, these direct external stressors have the greatest potential to have significant influence on wastewater volumes and concentrations.

THE EVOLUTION OF RESILIENCE AS A CONCEPT IN THE WATER SECTOR

Since the original research of Holling (1973) and Walker et al. (2004), interest in resilience has grown in all subject areas. The investigation of resilience as a determinant for water and wastewater systems has been well documented, see Butler et al. (2014). A reasonable level of success has been achieved in water supply resilience, although this is often not suitable for the complex multi-variate mechanisms, and often competing processes, associated with wastewater delivery and treatment. When wastewater is generated numerous factors are involved, e.g., urban creep or groundwater infiltration to sewers can dramatically increase the flow to a receiving WRRF. The randomness and unpredictability of changes leads to significant reductions in resilience over a short period of time (i.e., hrs). Attempts have been made to circumvent this with research focussing more toward general resilience (Sweetapple et al., 2022a), which utilises a more systems of systems approach that can be highly complex. From an industry perspective, there has been growing interest in resilience as a concept by the UK water companies; however, the principles have not been fully embedded. This was exemplified by the Ofwat price review 2019 (PR19), where only two water companies provided evidence of securing the long-term resilience of their assets (Ofwat, 2019). Therefore indicating that the challenge is not the generation of greater knowledge; but applying resilience theory and embedding its principles into daily operation.

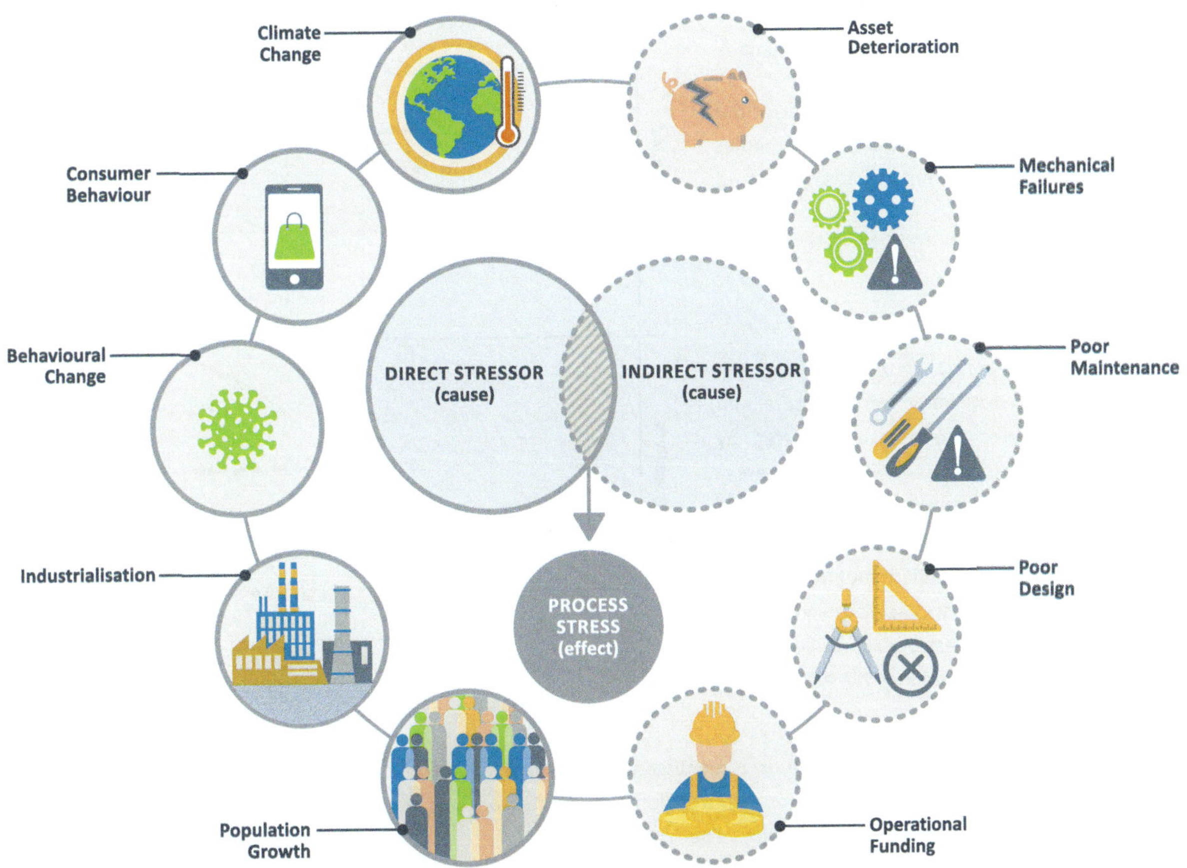

Figure 3.10 Direct (external) and indirect (internal) stressors adapted from Butler et al. (2014).

For this to happen, dynamic changes in operation that influence resilience should consider the use of existing WRRF data to build a knowledge base of past stressors and process stresses/resilience (dynamic resilience) in preparation to future events. For instance, to avoid potentially damaging pollution incidents, it is crucial to understand how stressors manifest leading up to such dramatic changes in wastewater volume and concentration. To understand these dynamic events, the interaction of stressors and process stresses/resilience should be considered as dynamic resilience (Figure 3.11). The principle of dynamic resilience in Figure 3.11 shows the stressor as having a bell-shaped peak and the process stresses generated as a well-defined peak concentration. Differences between stressors (cause) and process stresses (effect) occur because the WRRF behaves differently to the generated stressor due to recirculations and sludge extractions. Understanding the magnitude and duration of events has benefits, gives rise to the possibility of classifying stressor influence at the WRRF or for separate processes. It also allows for a reaction time between an event occurring (stressor) and its effect on the process (stress).

The definition of dynamic resilience used in this section, as shown in Figure 3.10 and also presented in the International Water Association (IWA) Modelling and Integrated Assessment (MIA) Specialist Group (SG) webinar (Holloway et al. 2021) is detailed below as:

"The dynamic, temporal variation of stressors and process stresses (and resilience) in response to events outside of standard operating conditions"

Aims and objectives

This section starts with expert perceptions of resilience, then approaches the challenge of evaluating the dynamic resilience in the context of actual WRRF data. It does this by demonstrating some common pitfalls of using actual WRRF data, then how stressors can be extracted for the evaluation of process stress/resilience effects. Dynamic resilience is demonstrated through a case study using 10 years of data to visualise a specific event. The broader impacts of dynamic resilience are also discussed, along with how methods could be incorporated to provide a holistic, resilience based approach which is applicable to water companies (Holloway et al., 2021).

Expert's perspectives on resilience

The following two sections provide an overview of the resilience based challenges faced by water companies and those interpreting data from wastewater-based epidemiology.

RESILIENCE FROM A WATER COMPANY PERSPECTIVE

Wastewater flow data is central to understanding the dynamic resilience of wastewater infrastructure and WRRF and to understand the pressure exerted by wastewater catchments (stressors). Water companies in the UK are making great strides in understanding how their assets react to climate change while incorporating meteorological and demographic data. In this section, Dr Ben Martin speaks of the challenges associated with asset and infrastructure resilience and how Thames Water are embedding dynamic digital systems.

Improving asset and infrastructure resilience is a significant challenge for the water industry as operational disruptions become more common and difficult to predict. The most significant of these disruptions are extreme weather events, which have recently delivered a month's worth of rain in 1 hour. Most sewers and wastewater treatment plants are many decades old and were simply not designed to cope with the loads and temperature fluctuations recorded in recent years. Such weather events have been coupled with the COVID pandemic, which resulted in spatial disruptions to normal wastewater loads as the mobility of the population reduced during lockdowns.

Thames Water is currently working on a host of data-driven projects to increase the resilience of ageing asset and infrastructure base. These digital initiatives will output control systems that can manage disruptive loads more effectively and efficiently. Additionally, a digital twin is being developed for Beckton sewage treatment works, the largest wastewater treatment plant in the UK. This brings together a number of hydraulic, pneumatic, and biological models to digitally represent its physical assets. This is expected to deliver a 10—20% reduction in workforce planning, a 30—50% reduction in predictive maintenance, and a 20—40% reduction in reactive maintenance.

A broad rollout of connected digital monitoring instruments throughout assets and infrastructure is planned. These include prediction systems so that assets can be maintained and replaced long before failure, machine learning to enable autonomous waste catchments, and open data frameworks

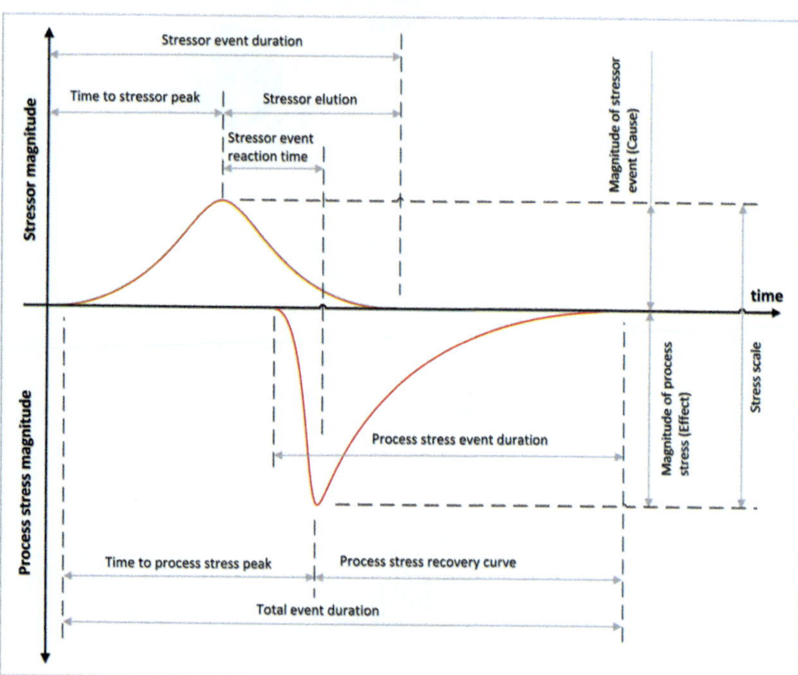

Figure 3.11 Dynamic resilience diagram showing the separation of stressors and process stresses.

for sharing across water companies. Linkages will also be established with meteorological and demographic datasets. These will inform operational control strategies that can respond in real time, or even ahead of time, to enable rapid recovery from events. These dynamic digital systems will ensure that as disruptions become the norm, wastewater can still be treated to exceed environmental targets.

RESILIENCE OF WATER MANAGEMENT SYSTEMS FOR PUBLIC HEALTH PROTECTION

Social resilience has a key role in the protection of public health. In this section, Dr Matthew Wade writes of the challenges associated with asset and infrastructure resilience and its impact on monitoring wastewater-based epidemiology (WBE).

Developments in the water sector to embrace Industry 4.0 and digital transformation philosophy have demonstrated an evolution in the function of wastewater. While the focus of treatment and transport infrastructure has been generally on protecting the environment from harmful pollutants and, more recently, resource and energy recovery from wastewater (Daigger, 2009; Guest et al., 2009; Kehrein et al., 2020), the COVID-19 pandemic has given greater visibility to a broader function of wastewater systems, public health protection.

Wastewater monitoring as a tool for public health intervention dates to the mid-19th Century, when physician John Snow mapped data of cholera incidence in Soho, London to determine the source of the outbreak (a water pump contaminated with sewage) (Tulchinsky, 2018). The detection of sewage borne indicators of public health, commonly known as WBE, has been used to monitor a range of targets from poliovirus to illicit drug use in urban centres (Larsen et al., 2021). From the onset of the COVID-19 pandemic, the SARS-CoV-2 viral RNA was shown to both be detectable and quantifiable in sewer samples collected throughout the sewer network, typically at the inlet of treatment works, within-network, or at near-source (e.g., building scale) (Ahmed et al., 2020; Sweetapple et al., 2022b).

Given the evolution of the virus, subsequent work has also demonstrated the ability to detect its variants (Crits-Christoph et al., 2021). Once evidenced, the challenge for those working with WBE for COVID-19 was to determine the value of these datasets for public health policy and decision-making. The rapid uplift of COVID-19 science in wastewater and the fragmented nature of its utilisation across the globe means that the true value proposition of WBE as tool to complement existing measures of public health remains unproven and uncertain. Factors influencing its resilience include a lack of empirical data to understand and mitigate for wide range of uncertainties associated with the data from WBE (Wade et al., 2022), a robust understanding of the relationship between the target marker(s) of public health (Mao et al., 2020), and the general lack of standardisation and protocols to enable WBE to be implemented as a function of public health policy (Wu et al. 2021). Nevertheless, WBE has great potential as a tool for public health protection beyond COVID and across a broad range of targets (e.g., lifestyle chemicals, pathogens, metabolites of health), settings (e.g., community-wide, critical infrastructure), environments (e.g., urban centres, low-income settings), and functions (e.g., rapid response, long-term surveillance of health trends, targeted monitoring).

For WBE to be useful in the long-term, its research, development, and use needs to be considered together with the wider efforts to ensure infrastructural and data resilience. This should be viewed as a resilience of the systems where information is collected, as a failing sewer network will inevitably lead to greater measurement uncertainty. Additionally, it is crucial to consider its ability to ensure public health resilience as the information acquired by WBE must be reliable and usable by the stakeholders receiving it. This effort must be global. Water and disease know no borders and ensuring the resilience of water systems in highly resourced regions (e.g., those with the means to embrace digital tools to manage the increasingly voluminous and valuable data streams), must be matched by initiatives to maximise the value and potential for WBE in low-income settings (e.g., development of low-cost but smart technologies (Gwenzi, 2022)).

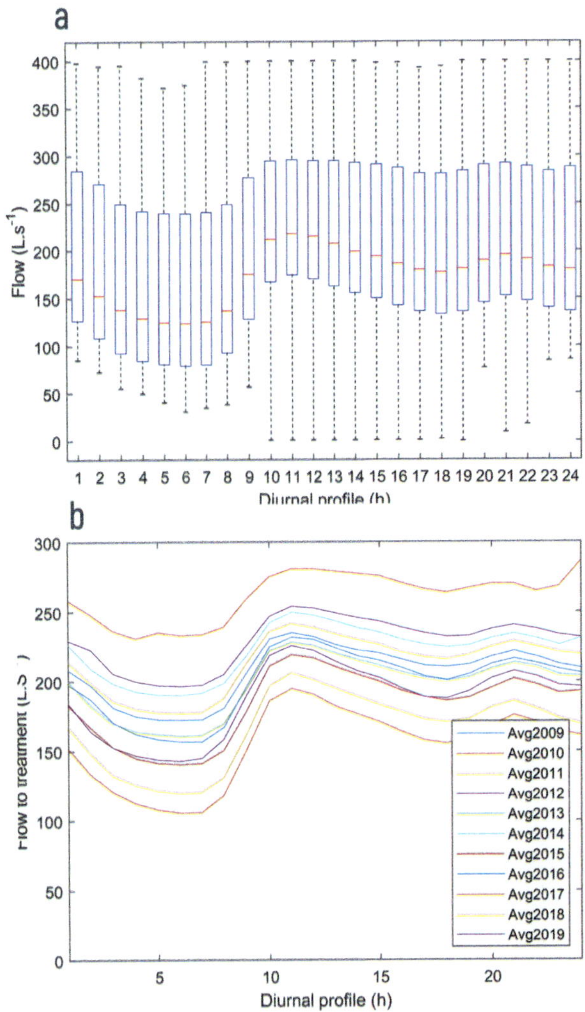

Figure 3.12 10 years of hourly flow data over a 24 h period aggregated into boxplots (a) and average flows by year (b).

The importance of data for the evaluation of dynamic resilience

This section evaluates the challenge of using actual stochastic data, presents possible methods for generating valuable insights, and finally evaluates dynamic resilience. It finishes with the future perspectives of dynamic resilience and how the best possible environmental outcomes can be achieved.

DATA AND THE IMPORTANCE OF SCALE: CHANGING FROM MICRO TO MACRO ANALYSIS

Before examples of dynamic resilience are presented, it is crucial to note some of the pitfalls of using actual WRRF data. The use of large datasets captured over multiple years can be overwhelming (e.g., 174,720 data points when logged at 15 min intervals over 5 years). Conventional concentration-time plots are not sufficient, providing poor resolution due to the variation frequency of actual stochastic WRRF data. An example of this can be seen in Figure 3.12a, where aggregated boxplots of hourly WRRF influent flow over 10 years show significant variability. It also indicates that this data is not normal or lognormal, so when averages are taken (Figure 3.12b), the central tendency is skewed providing inaccurate flow predictions. Therefore, significant uncertainty can be created when making statistical assumptions, particularly when data is being used as a timeseries for data-driven models or other statistical evaluations (Newhart et al., 2019). To avoid introducing significant uncertainty the entire data set must be considered (macro analysis). This also avoids the temptation to be miopic, analysing each discrete variation (micro analysis) which can also lead to erroneous outputs when performing simulations.

TOWARDS DIGITAL DYNAMIC RESILIENCE OF WRRF

The dynamics of digital systems have long been the interest of those working on solutions where real-world interfacing (sensors) is crucial for process control. Examples of this occur throughout robotic manufacturing and high-value fluids such as oil and gas. Unfortunately, being termed as a 'waste' rather than a 'product' places less importance on the value of the end product, resulting in less instrumentation (less concequential loss). Fortunately, for water companies, they generate data intensively from WRRF and processes. In decentralised rural locations, instruments are commonly used for compliance monitoring and are less likely to be used for real-time process control. However, larger centralised WRRF often combine monitoring of complicane with signals/data generated for process control. These larger WRRF processes have vast data silos spanning years or even decades. Much of this data generated for larger centralised WRRF holds clues to how dynamic stressors emerge and wheter they generate process stress/resilience. Therefore, to avoid exhaustive iterations for modelling and simulation of scenarios, this data could be used as a timeseries to evaluate the dynamic resilience and generate knowledge of past and present events through empirical and mechanistic modelling. Modelling improves the context of events, and the data permits evaluating them relative to actual conditions. Over time it may be possible to make predictions on how future stressors may influence specific WRRF processes. A data-driven dynamic resilience philosophy therefore uses actual WRRF data and established modelling practices to compute the impact of a stressor, then process related stresses/resilience. This takes the focus from the intensive simulation of event-based scenarios (stressors), to evaluating actual scenarios in the context WRRF instrument data.

Case study: dynamic resilience using 10 years of data from a WRRF in the south of the UK

In this section, 10 years of data has been used to (1) identify significant events through the evaluation of prominence and dominance, (2) extract a standard operating condition for an existing WRRF and (3) provide examples of dynamic resilience visualisations.

EVALUATING THE DOMINANCE AND PROMINENCE OF EVENTS UNDER DYNAMIC CONDITIONS

Events or stressors are typically characterised by their magnitude (the prominance of the variation). The most significant event magnitudes can be isolated then scaled for direct comparison. This can be done for both the stressor (cause) and process stresses/resilience (effect) which can be evaluated independently to isolate only the most severe, as shown in Figure 3.13. When a significant event has been isolated using prominence, the event dominance can be estimated as the difference between the stressor exerted to the WRRF and the resultant process stresses/resilience. The difference in time between the stressor and process stress peaks allows estimation of the reaction time for applying interventions (maximum time to react). It is then possible to classify events as stressors and process stresses to estimate a reaction time for future events (i.e., the time between the stressor peak and the process stress occurring). Without event insight, it may not possible to learn from interventions applied to WRRF that reduce process-related stresses.

Examples of prominence are shown in Figure 3.13a as time-based examples of stressor prominence (brown line) and process stresses prominence (red line). Significant events can be seen in Figure 3.13a, when stressors are close to 1 and the resulting process stresses approach -1. The event dominance is shown in Figure 13b (difference between the stressor and process stresses), allowing for isolation of events that have the most significant effect on process performance. Therefore, using prominence and dominance based analysis, stressor events can be isolated, then evaluated based on the effect the event has on the WRRF (process stress/resilience).

FAILURE AS A CONSEQUENCE OF NORMAL OPERATION

As much as we don't like to admit it, failure is a consequence of standard operation, originating from specific and operational changes or stressor influences. Therefore, as provided in the definition of resilience, failure also has a magnitude, with the extent of failure present over a scale. An example of this is failure to remove sludge from a system, which over time accumulates solids until a compliance breach occurs. This is far less significant, than a toxicity event that essentially kills off microbes in a biological system. Therefore, the extent of failure is crucial when evaluating resilience, meaning dynamic metrics then become scalable from a standard operating condition (where the process normally operates). Therefore, when we consider WRRF process failure, we must first consider the failure extent, and what constitutes a failure (*a failure of what and to what extent?*). For example, operational staff may consider the the color of the process fluids or visual tests to evaluate process conditions. These are empirical observations of indirect/internal stressors that prevent the process from functioning and are common. However, process scientists and engineers take more of a theoretical method of diagnostics, with failure defined as a compliance breach. Therefore, when considering events, it is essential to appreciate both empirical and theretical thought processes.

WRRF processes are subject to diurnal variation, which is also dynamic, but typical of standard operation for a specific wastewater catchment. Therefore, it differs from original WRRF process engineering design information, particularly when urban creep and populations in the wastewater catchment increase. The difference with dynamic resilience is that it takes the standard operation from actual operational data, rather than historial design information. This extracted condition is called the standard operating condition (SOC) relating specifically to the nuances of a catchment or WRRF. This can be done by using clustering methods to extract three flow conditions similar to that of Borzooei et al. (2020). The clustered outputs for a particular WRRF are shown in Figure 3.14a, and the extracted SOC in Figure 3.14b alongside actual time-based data. This SOC is important when considering WRRF that have been retrofitted with supplementary processes to increase WRRF capacity, which is common practice internationally. The extent of process failure is classified as the process stress index (PSI) and is anything outside a WRRF specific SOC.

The outputs of the clustering shown in Figure 3.14 take an engineered resilience stance (Holling, 1996) and must be further elaborated to include a safety factor or degree of freedom to exclude normal variation (Figure 3.14b). Also, outputs based on failure should consider the balance of compliance risk to operating costs, with evaluations based on the WRRF meeting permitted limits at the lowest operating cost. However, caution should be applied to a minimum operational cost approach so the process does not become unstable or vulnerable to changes resulting from external stressors. Therefore, evaluating the magnitude of variation (stressor) as the extent of failure allows

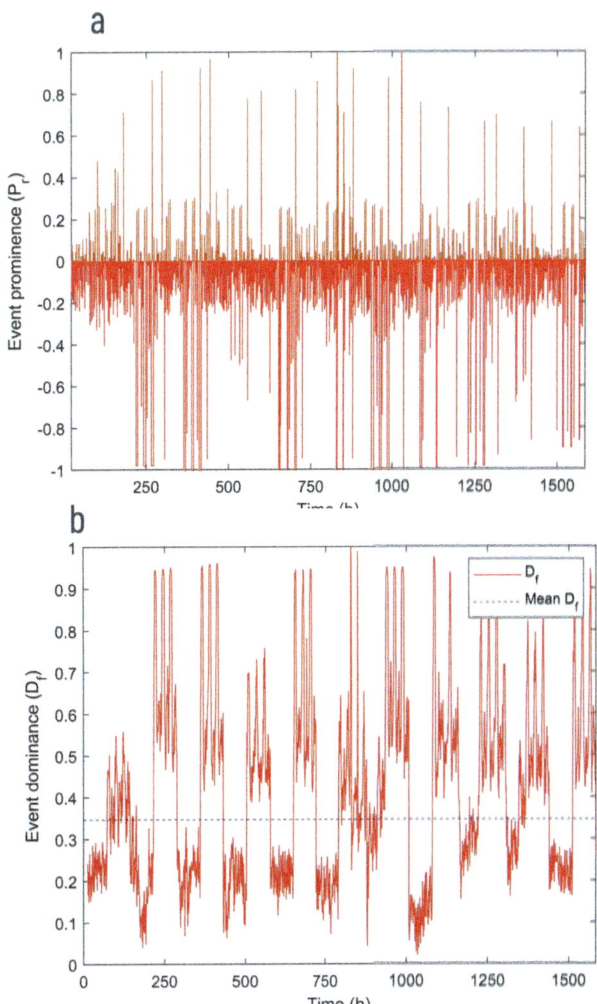

Figure 3.13 Prominence (a) and dominance (b) of stressors and process stresses adapted from Holloway et al., 2021.

failure prediction and numerous event classification possibilities based on process stress or resilience (i.e., dynamic resilience).

COMMUNICATION OF DYNAMIC RESILIENCE

Visual communication is often overlooked, which is possibly why Corominas et al. (2018) found that only 16% of publications on transforming data into knowledge reference a commercially available product. Therefore, it is possible that there is a distinct gap between the evaluation of water resilience and how it is communicated to those operating and maintaining wastewater processes. The communication of resilience data in the run up to a significant event can be challenging, particularly if there is no knowledge of past events and the associated interventions. Unfortunately, inadequate operational communication often comes as an incident investigation, such as the Longford gas plant explosion (Conlin and O'Meara, 2006). This highlights the

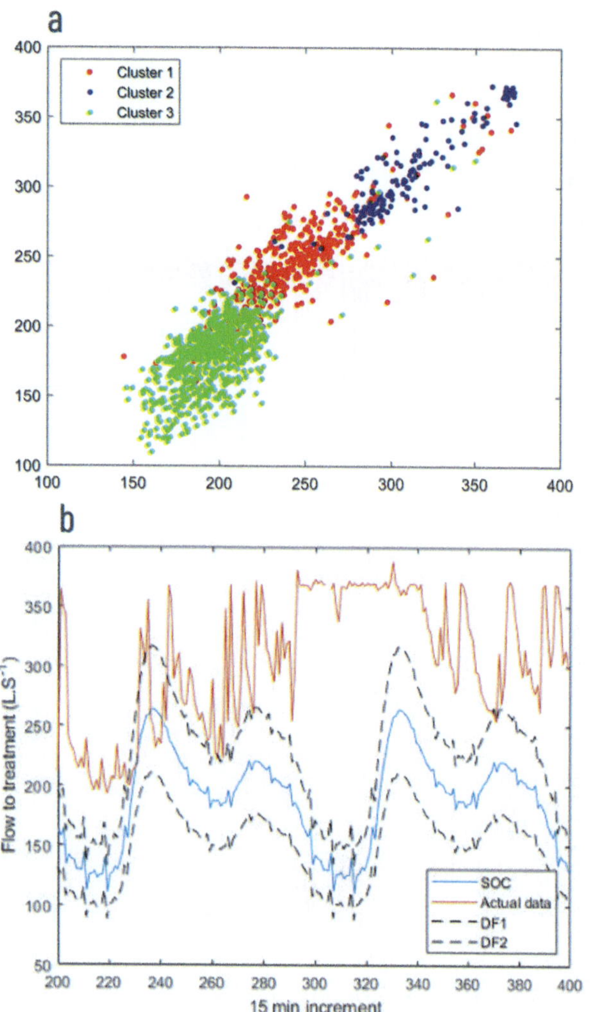

Figure 3.14 Examples of clustering methods for the generation (a) and outputted SOC including degrees for freedom (b).

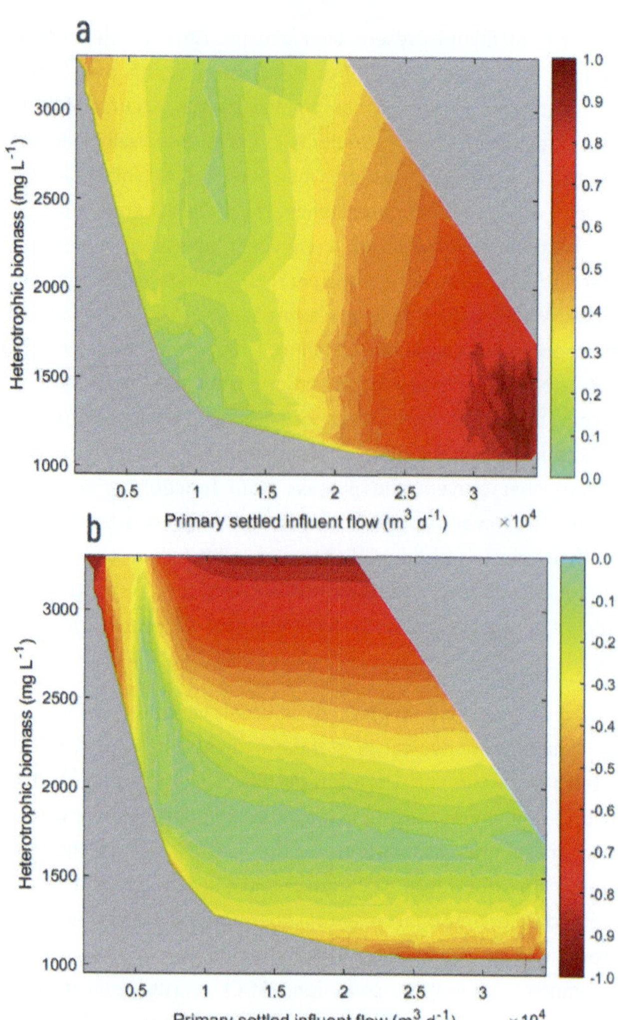

Figure 3.15 Influence of a stressor and generated process stresses using the SOW approach for activated sludge.

importance of linking resilience with complex modelled outputs in a communicable form for operational interpretation. It should also reflect the time-based dynamics of actual WRRF operating resilience.

To address this, the concept of dynamic resilience aims to incorporate time-based evaluations to visually communicate event severity to operational staff. self ordering windows (SOW) are used as a visualisation method for stressor and process stress/resilience observations. These SOW use a 48 hrs event window and, through transformation, plot the PSI as a contoured heat map. The SOW then becomes a significant event window based on its duration, prominence and dominance. In most cases, SOW represent a unique dynamic resilience fingerprint extracted from actual WRRF data, while keeping complex modelling practices out of sight (IWA ASM model series). The SOW principles also avoid having numerous number

of iterations required for methods such as Monte-Carlo and other iterative simulations.

Examples of dynamic resilience are shown in Figure 3.15 to demonstrate the impact of flow on heterotrophic biomass concentrations, where Figure 3.15a shows the stressor and Figure 3.15b the process stresses resulting from the stressor. The stressor in Figure 3.15a shows concentrated zones at the highest flows and lowest concentrations and process stress at the highest concentration and middle of the flow range. The grey areas in each plot indicates zones of no data. Therefore, the concept of data richness is also crucial to SOW success, where inputs must reflect the time-based range and specific variation of the WRRF.

The main challenge of dynamic resilience is the duration of the time-based window selected for evaluations. It can also be extremely difficult to predict when an event starts and ends, with characteristics varying depending on specific or multiple stressors. This will be crucial for the expansion of dynamic resilience to incorporate real-time control (RTC) for live management of stressors and process stresses/resilience generated at the WRRF. An extension of that could be a traffic light system, as shown in Figure 3.16, mapping the process stresses, but also communicating them for the application automated interventions.

Summary of dynamic resilience methods

The dynamic resilience approach proposed in this section has demonstrated the possibility of using actual WRRF data to understand and communicate dynamic resilience. The methods presented used prominence and dominance to isolate significant events, then a macro data analysis approach to extract a dynamic SOC. Actual data points were used to scale system failure magnitude and compute the dynamic resilience of a specific WRRF as a SOW, which reflects specific process nuances in response to significant events. This includes the possibility of isolating novel events generated by climate or societal change. However, the main challenge for dynamic resilience is selecting a suitable time over which dynamic resilience is monitored, as this can dramatically influence SOW output and affect the classification of events.

Overall, the dynamic resilience methodology has provided a possible link between resilience, data-driven modelling and visualisations through time based contoured heat plots (SOW). It is hoped that these methods could eventually close gap between evaluating and modelling resilience and its communication to wastewater operators within water companies globally. However, the methods presented are limited to countries that (1) have the instrumentation installed in WRRF or networks and (2) have the capacity and knowledge to maintain these instruments.

Reflection on dynamic resilience for an uncertain future

The future of the planet is reliant on how resilient the human race can be to changes in the climate and rapidly emerging stressors. As we face increasing uncertainty from political, social, and environmental factors, water companies and government agencies are forced to manage the dynamics of resilience resulting from changes outside of their control. Although many theoretical methodologies of resilience have been proposed, a unified approach has not yet been developed to (1) satisfy the dynamics of resilience that occurs from an actual WRRF and (2) communicate outputs to operational and maintenance staff.

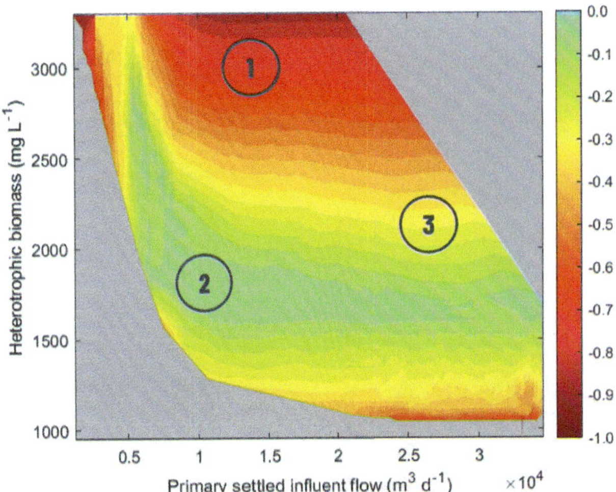

Figure 3.16 Process stress analysis using the SOW principles, adapted from Holloway et al., 2021.

At the time of writing, the Russian invasion of Ukraine is causing significant political and economic instability. This, again, will likely generate novel stressors that result in process stresses in wastewater assets and infrastructure as transient populations exit Ukraine to neighbouring countries. These countries will now be subject to increased demand for clean water and increased capacity to treat additional wastewaters generated by those that have exited Ukraine. It is extremely likely (Pörtner et al., 2022) that we will continue to see the emergence of novel, rapidly emerging stressors, and if we continue on the same path, there is high confidence that their occurrences will increase.

It is also important to consider the factors that have contributed to stressors in recent history. Reflecting on the past 3 years, the following factors/stressors have emerged globally:

1. **February 2022**: IPCC WGII sixth assessment report predicts with high confidence global increases in inland flooding, flood and storm damage in coastal areas and damages to infrastructure (Pörtner et al., 2022).

2. **February 2022**: the Russian invasion of Ukraine caused migration into neighbouring European countries.

3. **March 2020 to present**: the COVID-19 pandemic escalates, causing unprecedented damage to human health, global economies and freedom of movement (Ramos and Hynes, 2020).

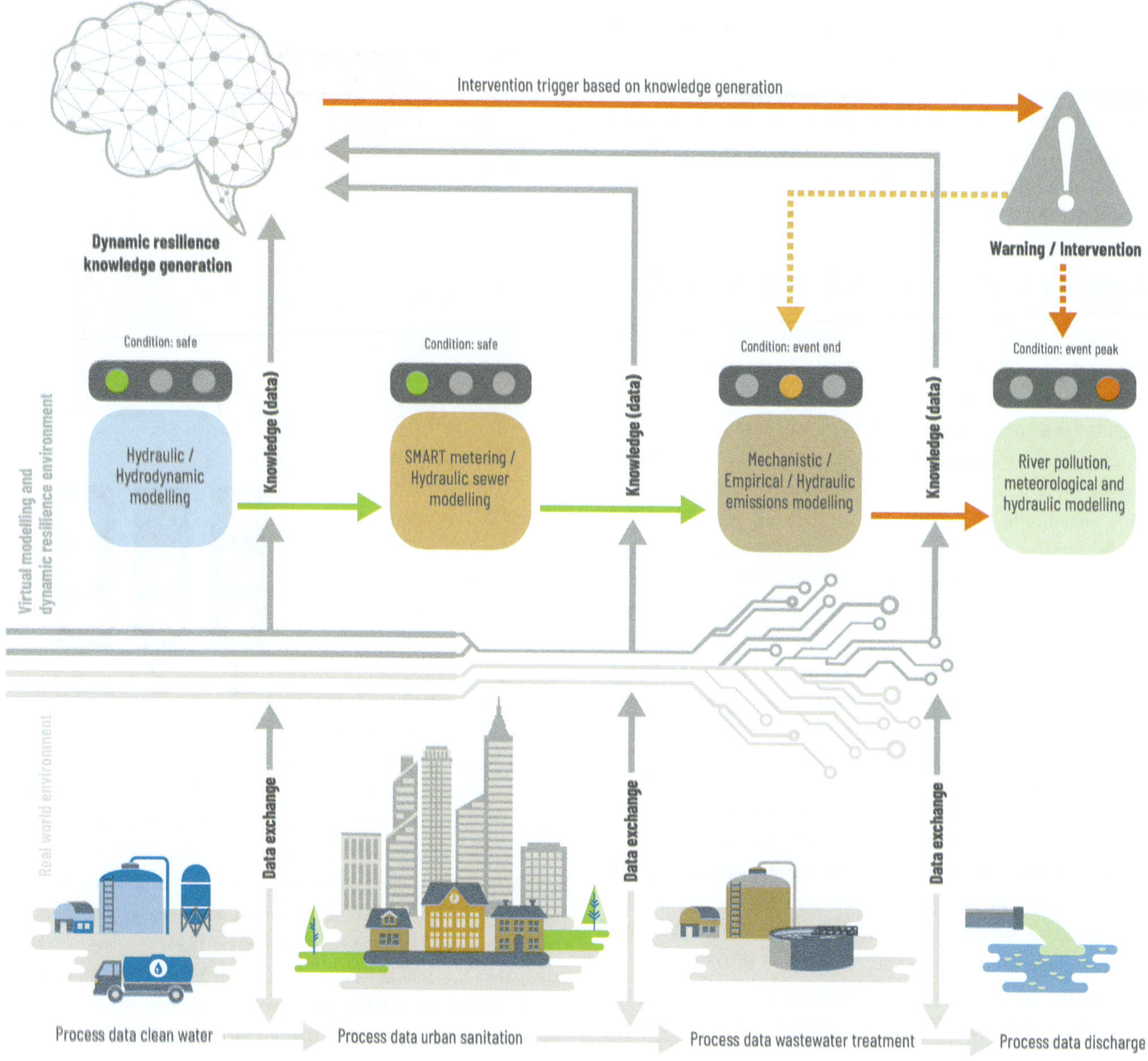

Figure 3.17 Dynamic resilience for autonomy of the urban water cycle under the precept of Industry 4.0.

The above emphasises the absolute need to understand the dynamic resilience of not just in assets and infrastructure but also due to societal change. Dynamic resilience has the premise of providing the means to understand the real-time adaptive capacity by monitoring stressors and process stresses/resilience directly from WRRF and associated processes (instruments). We must consider how existing instruments and data from these instruments can drive us toward Industry 4.0. Instrumentation combined with an understanding of dynamic resilience could allow us to understand and to adapt resilience in response to novel global events, while understanding where improvements can be made.

We already have many connected networks for intensive data exchange, along with large data silos (historical knowledge). Using these tools, it may be possible to access the real-time dynamic resilience of not just WRRF, but any interconnected infrastructure network. Therefore, the future of dynamic resilience and dynamic processes should embrace the possibility of improving resilience through digitally connected assets and infrastructure as the beating heart of the modern world (Figure 3.17).

Conclusion
Transforming information to insight

As shown by the case studies in this chapter, AI, DT and DR are three powerful tools that can be used by the international water industry to take advantages of 'big data'. These are also tools that water professionals now have at their fingertips. Most utilities and system designers/engineers have more data available than they can readily integrate and analyse efficiently. So any analytical tool than can rapidly extract valuable insights from existing data has the potential to reduce both operating and capital costs and also understand the resilience to external stressors such as climate change.

Artificial intelligence is now being used in the global water industry to find hidden patterns in large data streams and recommend the best approach to meet specific performance targets. As discussed above, the pattern recognition capabilities of AI offer the rapid identification of adverse conditions in water/wastewater assets, such as pipes and pumps. Additionally, AI has been used to recommend control actions that achieve objectives such as lower energy usage, predicting future events, and the identification of water/wastewater system data patterns. The ability of AI to identify patterns in large data sets with many variables allows additional value to be extracted from generated data. Instead of simply investigating trended data, many dimensions and variables can be investigated to understand any hidden characteristics within the data.

Many global industry challenges are driving the emergence of digital twins in the water/wastewater industry. These challenges relate to: (1) staffing, (2) the need to improve efficiency in water/wastewater related systems, and (3) the need to make more efficient decisions in operations. A digital twin of the system answers most of these drivers. Digital twins force the reconciliation of many data streams into a coherent platform that enables operations and maintenance staff to see the operational conditions more clearly to make more effective and efficient decisions.

Resilience is rapidly becoming one of the most critical water/wastewater asset-related challenges of this century and is aligned with how we '*do more with less*' where assets are commonly retrofitted rather than replaced. The emergence of climate change along with extreme modifications to human behaviour caused by the COVID-19 pandemic, which changed the volume and composition of wastewaters, has further compounded these asset-related challenges. These stressors have reduced the resilience of our water/wastewater assets and infrastructure increasing their vulnerability and the potential for pollution events. To address the the impact of these stressors, it is crucial to understand that resilience is dynamic and may vary diurnally, monthly, seasonally, and even in response to the external events and stressors. The dynamic resilience methodology presented in this chapter uses actual WRRF data, machine learning, data-driven modelling, and heat map visualisations to capture the apparent stresses. These visualisations can be used to evaluate the stress response of a process to an extreme event or investigate its prominence/dominance and the possibility of the process failing (pollution event or reduction in process performance). This could lead to the prediction and anticipation of extreme events and the rapid deployment of event-based mitigations.

The combination of AI, DT, and DR could entirely change how the water/wastewater industry operates, maintains, and understands its facilities. The pattern recognition capabilities of AI can be used to augment DT to free up operations staff for critical time relevant decision-making and not get stuck in the mire of digging through large data sets (micro analysis). Also, the relevance of understanding and evaluating dynamic resilience from actual data can support water/wastewater operational decisions by relating them to the resilience or reserve capacity of a WRRF system. This combination of analytical tools also opens up possibilities for making more complex (and more efficient) systems designs (biological, chemical, and physical) for a broader range of utilities that may not be able to recruit and retain the expert workforce typically required for complex operational systems.

Chapter 4

Digital transformation in practice

Introduction

In this book so far, we have seen some of the tools that help us achieve digital transformation from the building blocks of instrumentation and the data it brings, the information behind that data – the meta-data — and how to manage that data through uncertainty analysis. Building on this are the techniques to use this data from artificial intelligence and machine learning through to modelling techniques with instrumentation (i.e., Digital Twins) and the use dynamic resilience using techniques such as multi-variate process control.

All of these tools and techniques are just a selection of what it takes to help the global water industry achieve digital transformation. However, what does this look like in practice? Some of these techniques sound very complicated and very expensive and not very accessible. In reality though, nothing could be further from the truth and Digital Transformation can be applied very economically from small water systems all the way to company-wide systems. It of course depends on the cost-benefit of the digital transformation solution.

In this chapter will look at some of the applications of the digital transformation in public health and smart sanitation, its use in wastewater-based epidemiology (WBE) in order to inform public health, and in sewer blockage detection to assist in pollution prevention and environmental protection.

4.1 Improving public health through smart sanitation and digital water

Digitalisation is a tool that connects assets, allows for more situational awareness and a greater understanding of what is happening, and brings together data for informed decision-making. To cope with the increasing difficulties in managing scarce water resources, the future calls for a digital water and sanitation economy with a smart approach to design, use, application, and control.

Collaborative ecosystems

As sanitation and water systems become more digital, a distinct opportunity arises to share these data streams with municipalities for the benefit of the consumer and the city. Increasingly, we are seeing sanitation and water operators work with municipalities and governments to map their systems and usage data. With emerging technologies, we can envision public health and infectious disease or vaccine monitoring layering on top of that data.

Technological advances around preventative care and precision medicine are converging and moving us to greater and more patient-centric health systems. Sanitation and water systems can provide routine data collection and health monitoring tools, all whilst creating new value for the existing system and shifting responses to disease outbreak from reactive to preventative. As evident during the COVID-19 pandemic, when it comes to disease, early detection, treatment, and community response are all vital to limiting the impact on an individual's health and preventing outbreaks within a community. Through equipping sanitation and water systems with smart sensors, they can begin to detect disease and monitor environmental and societal transmission factors quickly and autonomously. The data collected can be used to provide evidence-based decision support for those tasked with managing the spread of disease, while predictive health analytics can assist early-warning systems.

A global data architecture that facilitates the rapid and efficient sharing of data and information from countries, states, or territories is a key component of global surveillance. While the International Health Regulations stipulate the legal responsibilities to inform the World Health Organization (WHO) about the occurrence of certain public health events, there is currently no harmonised public health reporting mechanism that enables information exchange from public health institutes and agencies directly to WHO. Such a mechanism is needed to access and analyse the disaggregated data required to understand age- and sex-specific epidemiological features, risk characteristics of certain sub-groups, and distributions of cases over time and geographical areas.

The foundations of such an architecture have already been laid through the creation of the Epidemic Intelligence from Open Sources (EIOS) data platform, which allows multiple communities of users collaboratively to assess and share information about outbreak events in real-time. The future vision of the new data architecture has been articulated by the Epidemic Big Data Resource and Analytics Innovation Network (EPI-BRAIN) initiative, which harnesses cutting-edge tools for big data, crowdsourcing, and artificial intelligence. This mitigates the impact of epidemics by allowing stakeholders to merge public health data with other data sets that can provide insights into the complex factors that drive epidemics. This includes data on human and animal population movement, animal diseases, and environmental and meteorological factors. Using advances in language processing and machine learning, these data sets are used to provide a more comprehensive analysis that helps to predict outbreaks and track their spread.

A unique role for business

By proactively applying sanitation economy approaches, companies have the potential to break down significant barriers and positively impact the lives of millions of the world's most vulnerable. Scaling up new sanitation economy approaches within businesses will ensure sanitation access and unlock resources and data that will transform the economics of sanitation into commercially viable opportunities.

In light of the COVID-19 pandemic, WHO has recommended that workplace preventative measures are established to reduce risk, including the appropriate directives and capacities to promote and enable standard COVID-19 prevention measures in terms of physical distancing, hand washing, respiratory etiquette, and, potentially, temperature monitoring. Additionally, WHO advocates that communities must remain fully engaged in detecting and isolating all cases, that behavioural prevention measures must be maintained, and that all individuals have key roles in enabling and, in some cases, implementing new control measures.

After COVID-19, we anticipate global companies taking greater interest in the health of their employees in the workplace. Already we see temperature checks, frequent testing, and other measures being rolled out across operations. Smart sanitation systems are also an opportunity for businesses to ensure a safe and healthy work environment for their employees.

CASE STUDY 1: THE CASE OF KAMPALA

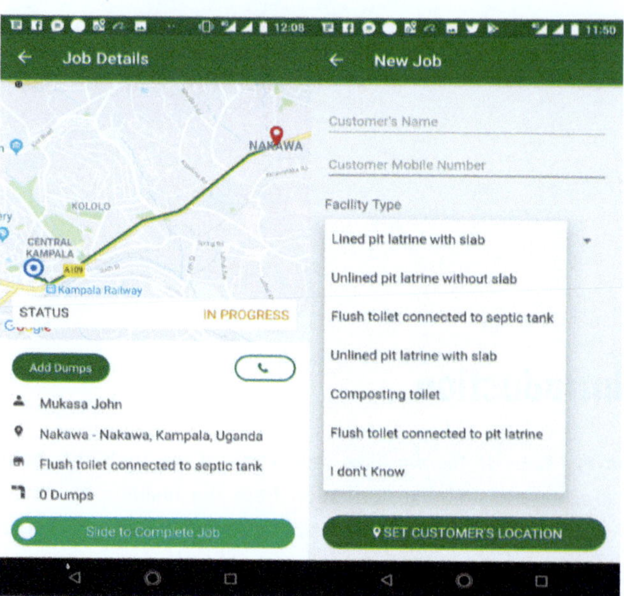

Figure 4.1 Sample GIS mobile application interface courtesy of GSMA, 2020

In Kampala, Uganda's capital, over 60% of the population lives in informal housing, while only 10–15% of the city is connected to formal sewerage. In this context, pit latrines and septic tanks are often emptied haphazardly by independent pit emptiers, who may dump waste illegally into the environment.

In response to the challenge of delivering sanitation services to Kampala's urban poor, the Kampala Capital City Authority (KCCA) launched a geographic information system (GIS)-based mobile application that links pit emptiers with customers (Figure 4.1). A grant from the Global System for Mobile Communications (GSMA) M4D Utilities Innovation Fund in 2017 was used to upgrade the pilot GIS tracking system, build capacity, and promote pit- emptying businesses. KCCA receives pit emptying jobs from customers through its call centre, connecting customers with the nearest pit emptiers. Pit emptiers submit critical data through the app to KCCA, including customer details, the amount paid, volume emptied, and the type and location of the sanitation facility. The app and call centre serve as an "ecosystem catalyst" by connecting customers with sanitation services and helping to ensure safe disposal of faecal sludge for a cleaner and healthier city.

The platform also enables KCCA to map sanitation activities across the city, which allows them to monitor and regulate service delivery and identify locations in need of interventions. As of January 2020, 171,000 sanitation facilities, such as pit latrines, were mapped. Insights from its geodatabase and sanitation customer call centre have provided KCCA with actionable information, such as the characteristics of sanitation facilities, how frequently pits are emptied in different districts, and the distances between pits and waste treatment plants. Given that 30% of all pit latrines in Kampala's informal settlements are still emptied into the environment, KCCA aims to use this information to target and guide investment planning, allocate resources, and regulate service delivery and standards enforcement. KCCA also leads targeted advertisement campaigns to increase demand for pit emptying in districts where health outbreaks are recurrent and relate it to the reduction in disease outbreaks. According to surveyed users, there has been a reduction of 87% in illicit disposal of faecal sludge in the communities.

KCCA has also invested in supporting and empowering the pit emptiers in using digital tools, such as a mobile application. The app facilitated over 5,000 pit emptying jobs, improving overall sanitation in the city, and building the capacity of pit-emptying entrepreneurs. Those using the app reported a 63% increase in income, with 85% of pit emptiers reporting using the app regularly.

Using digital tools to inform responses to urban sanitation in the public sector is not only important in the short run but can also contribute to building long-term digital capacity in the public sector, which can be crucial to long-term sustainable service delivery and the city's ability to respond to sudden shocks, such as pandemics or natural disasters.

CASE STUDY 2: GATHER

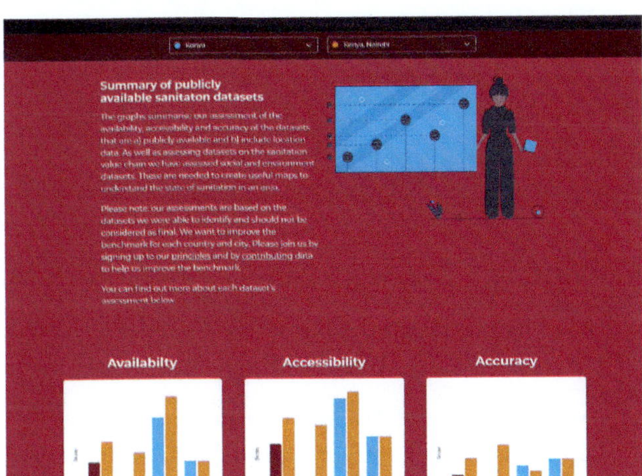

Figure 4.2 Gather Benchmark Tool to improve data integrity, 2020

Gather is transforming how sanitation data are collected, shared, and used by organisations working in Antananarivo, Republic of Madagascar. Gather's approach harnesses geospatial data to help municipal decision-makers understand where and how

to invest to improve sanitation infrastructure and services for vulnerable communities.

Gather's Centre for Sanitation Analytics recently completed the first assessment of the sanitation data gap in Antananarivo, to help decision-makers prioritise the collection of location data on the sanitation value chain. Gather have also led the creation of a new data standard, sanitation risk index, and sanitation data platform in partnership with Antananarivo's Commune Urbaine d'Antananarivo, Service Autonome de Maintenance de la Ville, Water and Sanitation for the Urban Poor, and Loowatt SARL. These three data tools are being refined, tested, and improved to transform how organisations collect, share, and analyse sanitation data. The aim for the geospatial visualisations is to make it easier for organisations to identify priority areas for investment for 350,000 people.

Gather have also launched a new Benchmark tool to help improve the integrity of location data collection for sanitation and development for the 50 countries at the end of the UN Human Development Index (Figure 4.2). The tool allows decision-makers to explore and compare the availability, accessibility, and accuracy of publicly available sanitation data, including location; it encourages the global sanitation sector to sign up to eight principles to overcome some of the practical and ethical problems with how data are currently collected, shared, and used; and it invites local decision-makers to share publicly-available data so that the tool's analysis can be updated.

CASE STUDY 3: THE UNDERWORLDS PROJECT

The Underworlds project was launched in 2015 to develop a human health census by sampling the "urban gut" at multiple locations, thereby increasing the spatial and temporal resolution of sewage sampling and analysis. Pioneered by the Senseable City Lab and the Alm Lab at the Massachusetts Institute of Technology (MIT), in collaboration with other laboratories within MIT and sponsored by the Kuwait Foundation for the Advancement of Sciences, Underworlds combined automatic samplers, biochemical measurement technologies, data visualisation, and downstream computational tools and analytics.

Rather than sampling sewage downstream at the wastewater treatment plant, Underworlds focused on upstream sampling throughout the sewage network in the city to develop individual readings of neighbourhoods. The Underworlds project gave rise to continued disease-monitoring work in the Alm Lab, including typhoid and, more recently, COVID-19, and to the creation of Biobot Analytics, Inc., which spun out of the Alm Lab in 2017 to provide wastewater-based data to combat the opioid epidemic in the US, and which is currently engaged in widescale COVID-19 surveillance.

Digital Water

Water and sanitation are among the most essential services a city provides and are at the foundation of economic stability. As stated in the IWA white paper Digital Water: Industry Leaders Chart the Transformation Journey,

> "The emerging smart city initiatives are creating demand for digitalisation across industries as cities face new challenges connected to the need to optimise infrastructure, industries and services. In this context, the very nature of a water (and wastewater) utility can be used as a springboard to a 'Smart City' by increasing connectivity and better engagement of governments, citizens and businesses. In developed countries, water utilities touch virtually every citizen, home and business, meaning cities can exploit a digital water utility's communication network, customer base and immediate value propositions to demonstrate and communicate the overall benefits and successes of the city being an interconnected enterprise."

Digitalisation in the water sector is rapidly growing in developing countries. Furthermore, as countries strive to achieve the Sustainable Development Goals in the coming decade, new markets are emerging across Asia, Africa, Latin America, and the Middle East for digital technology to meet increasing demands for the challenges in dense, urban areas and water-scarce regions. Some of these challenges were reflected on by George Bauer at GSMA in a blog written for IWA and include the following:

- **High non-revenue water:** A key constraint to more inclusive water service delivery among utilities in developing countries is non-revenue water (NRW), which is water that has been produced but 'lost' (due to leakages, faulty meters, illegal connections or non-payment) before it reaches the customer. These commercial losses alone are estimated to cost utilities in developing countries $3 billion per year.

- **Ineffective billing collections:** The cost of cash collections for utilities in developing countries can range from 3 – 20 percent of total revenue, severely restricting their ability to extend their reach to poorer customers and invest in network maintenance and innovation.

- **Disconnection/Non-connection of poor households:** In sub-Saharan Africa, just under 25% of urban households have access to piped water. For those living in crowded and unplanned informal settlements, the physical layout and land tenure issues may prevent them from connecting. Another barrier is lump-sum, unpredictable bills that may come irregularly and overwhelm a poor household's ability to pay. Water utilities' economics depend on having as many paid connections as possible to pay for maintenance and keep the service affordable for all, so not connecting poor households leaves out part of their market.

- **Insufficient investments:** Investments in urban water supply, in particular, have not kept up with urbanisation and population growth. This explains why, according to the World Bank, the proportion of urban residents with access to safely managed drinking water in Sub-Saharan Africa has barely increased over the past 15 years.

Digital tools and processes can be tailored for different areas of the water value chain, including in developing countries, e.g., using sensors, digital payments, and/or mobile phones. Real applications of digital tools and the solutions, as well as the benefits they provide, are outlined in Box 1.

BOX 1

EXAMPLES OF APPLICATIONS AND ADVANTAGES OF DIGITALISATION

Some examples provided by George Bauer at GSMA from his IWA blog include the following:

- **"Digitalisation of utility processes:** Digitising processes, such as meter reading, billing, payments, and complaint management systems, have shown a clear reduction in NRW for many utilities. For instance, Wonderkid, a Kenyan company offering Integrated Mobile Utility Management software systems to water utilities supported by the M4D Utilities Innovation Fund, developed a digitalised complaint management system for water utilities in Kenya. By fostering transparency, traceability, and accountability, IMUM makes water utilities more accountable to their customers, which in turn increases their willingness to pay.

- **"Pay-as-you-go water:** For low- income customers, it is easier to pay for what they consume in smaller amounts rather than a lump sum at the end of a billing cycle, especially for a service that may not necessarily have provided a steady, timely and safe supply of water. Using mobile payments also saves customers time and money by providing a secure channel to pay for water at a fair and set price without the need to travel to a local utility office. In September 2015, the GSMA M4D Utilities Innovation Fund awarded CityTaps a grant to launch smart prepaid water meters in Niamey, Niger, in partnership with the local water utility, Société d'Exploitation des Eaux du Niger (SEEN) and Orange Niger. Following the grant, customers reported savings of up to 95% of their spending on water from \$3.37 per m^3 to \$0.21 per m^3, and a reduction of 86 minutes spent collecting water on a daily basis (from over 90 minutes to under five minutes). Driving digital payment options among customers can have important benefits for water utilities and their customer base. Digital payments can reduce operational expenses and streamline service delivery. GSMA surveyed 25 water service providers (both centralised utilities and decentralised water service providers), and identified the following benefits associated with introducing digital payments for water services:

• Digital payments reduce collection costs by 57-95 percent;

• Digital payments increase revenues (between 15 and 37 percent) and enable new business models; and

• Digital payments increase customer reach by allowing utility service providers to operate at a greater scale and lower cost per customer.

"Though it is encouraging that there are more and more utilities scaling and piloting digital payment solutions, there is still tremendous scope for growth Combining with smart meters which record customer water usage can provide a clear picture of water consumption and convey data to both consumer and utility, allowing for improved water management, water savings and reduced costs."

An additional example extracted from Digital Water: Industry Leaders Chart the Transformation Journey, focuses on sensors:

"Sensors for real-time data: Sensors are a key component of digitalisation of water utilities as they can provide near real-time data on water quality, flows, pressures, and water levels, among other parameters. A variety of sensors, both fixed and mobile, can be dispersed throughout systems to aid daily operations by optimising resource use (e.g., chemical use for water treatment), detect, diagnose and proactively prevent detrimental events (e.g., pipe bursts, water discoloration events, sewer collapses/blockages, etc.), and provide useful information for preventative maintenance and improved long term planning for water utilities (e.g., by helping to prioritise repairs and replacements for aging infrastructure).

Sensors can also provide evidence for pipe corrosion and alert homeowners and utilities when water quality standards are not being met."

The why and how of digitalisation: the five-stage digital water adoption curve lays down a process that ranges from immature digital development phase to sophisticated use of technologies (not- developed, basic, opportunistic, systematic, transformational.) As stressed by the IWA network, the question of why the digitalisation of water utilities is critically needed is clear: the value of lowering operational expenses and, in turn, revenue requirements, has been proved for most industries. The more important question is how to implement and execute digitalisation in the water industry. This process is especially difficult in the water sector, which in most countries is regulated, and involves multiple different stakeholders and their underlying interests. For utilities operating in developing countries in Africa and Asia, it is even more complex as losses from operational inefficiencies are greater and resources to invest in digitalisation sparser.

The digitalisation process raises several questions; the first is, "where to start?". A utility must decide which departments or processes to prioritise for digitalisation. To understand the level of digital maturity in the water sector, using insights from 40 utilities worldwide the IWA white paper "Digital Water: Industry Leaders Chart the Transformation Journey" presented a five- phase "digital water adoption curve", which is both an assessment tool and a roadmap to guide utilities in their digital adoption journey. It begins with utilities that are in an immature digital development phase and need actions such as a strong push from top management to recognise digitalisation as a priority and pilot projects to explore opportunities to move forwards. As utilities move along the adoption curve, they must increasingly align themselves with data-driven goals and ensure that their processes evolve with changing technology requirements. Introducing digital tools and embedding these in the culture and strategy enable progress towards a "transformational" stage of adoption.

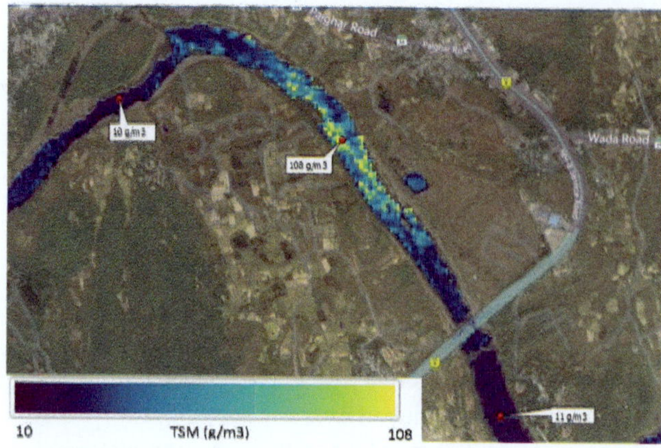

Figure 4.3 Satsense Solutions Vaitarna River, 2020

CASE STUDY 4 SATSENSE SOLUTIONS

Satsense Solutions uses satellite remote sensing and geospatial analytics to provide business and governance solutions (Figure 4.3). In a project co-funded by the European Space Agency and supported by the Toilet Board Coalition, Satsense Solutions used satellite remote sensing and artificial intelligence to develop a service to monitor effluence and water quality parameters in surface waters. This service has been used successfully by water utilities and water managers in the water, sanitation, and hygiene sector to assess the condition of water bodies and identify the source and intensity of pollutants in surface waters. The prevalence of untreated sewage and pollutants in surface waters could be correlated with risks of water-borne diseases.

Once wastewater-based epidemiology (WBE) processes detect the presence of pathogens in a community through sewage surveillance, geospatial analytics may be used to trace the extent and reach of the sewer system and highlight the affected community on a map. This may then be demarcated as a high-risk area and adequate measures can be taken by public health officials to address infections in the community.

Additionally, satellite remote sensing data might be used to estimate risks of infections within communities, on the basis of connectivity (such as road and rail connectivity, shared water resources, etc.), contamination pathways (water, human contact, etc.), population density (common healthcare facilities, schools, etc.) and transmission dynamics (vector-borne, human transmission, etc.). Such methods have been researched and used previously by NASA (the National Aeronautics and Space Administration) in estimating the risk of vector-borne disease and respiratory infections in communities using satellite data. This includes the risk of mosquito-borne diseases such as malaria and dengue, and water-borne diseases such as cholera. These methods and tools could play an important role in highlighting the risks of diseases and providing an impetus for improving sanitation in these communities.

The sanitation economy

Sanitation is one of the most pervasive yet overlooked development challenges facing us in the 21st century. An estimated 2.3 billion people around the world still lack access to basic sanitation, and 4.5 billion people — more than half the world's population — still lack access to safely-managed sanitation along the entire service chain. Increasing urbanisation is aggravating sanitation issues, environmental degradation and public health.

Sanitation as a business is growing and has the potential to unleash innovation, economic growth, and development with speed and scale. The Business and Sustainable Development Commission's "Better Business, Better World" report (2017) places water and sanitation infrastructure in cities among the 60 biggest market opportunities related to delivering the Global Goals. Furthermore, its value could be worth at least US$12 trillion a year in market opportunities and generate up to 380 million new jobs by 2030, with more than half this value in developing countries.

Sanitation is no longer a sector associated with the 'yuck' factor; instead, it is seen as a crucial opportunity for development projects and an opportunity for businesses to transform behaviours. The sanitation economy is a robust marketplace of products and services, renewable resource flows, data, and information that could transform future cities, communities and businesses.

Smart sanitation in cities

Digital technologies and data are opening new ways to re-think sanitation services in cities. A key underlying principle of the sanitation economy is that sanitation is not a system apart, but an integral and visible part of the wider infrastructure, services and resource flows. Many cities lack reliable information about public and community toilet usage, the quality of wastewater and sewage running through the system and the spread of infectious diseases. Sanitation can be included in the architecture of smart cities through data monitoring of public and community toilets. Technology can provide unique opportunities for municipal authorities to retain regulatory oversight of their cities' sanitation services while collaborating with business and investment to build, refurbish, operate and maintain the system. Monitoring and analysing this data will provide more efficient and effective public sanitation services in the world's growing cities.

Mapping and monitoring toilets, waste treatment, and health through control centres in smart cities creates new sanitation intelligence and a new lens through which to translate citizen and city sanitation needs into targeted solutions and new value

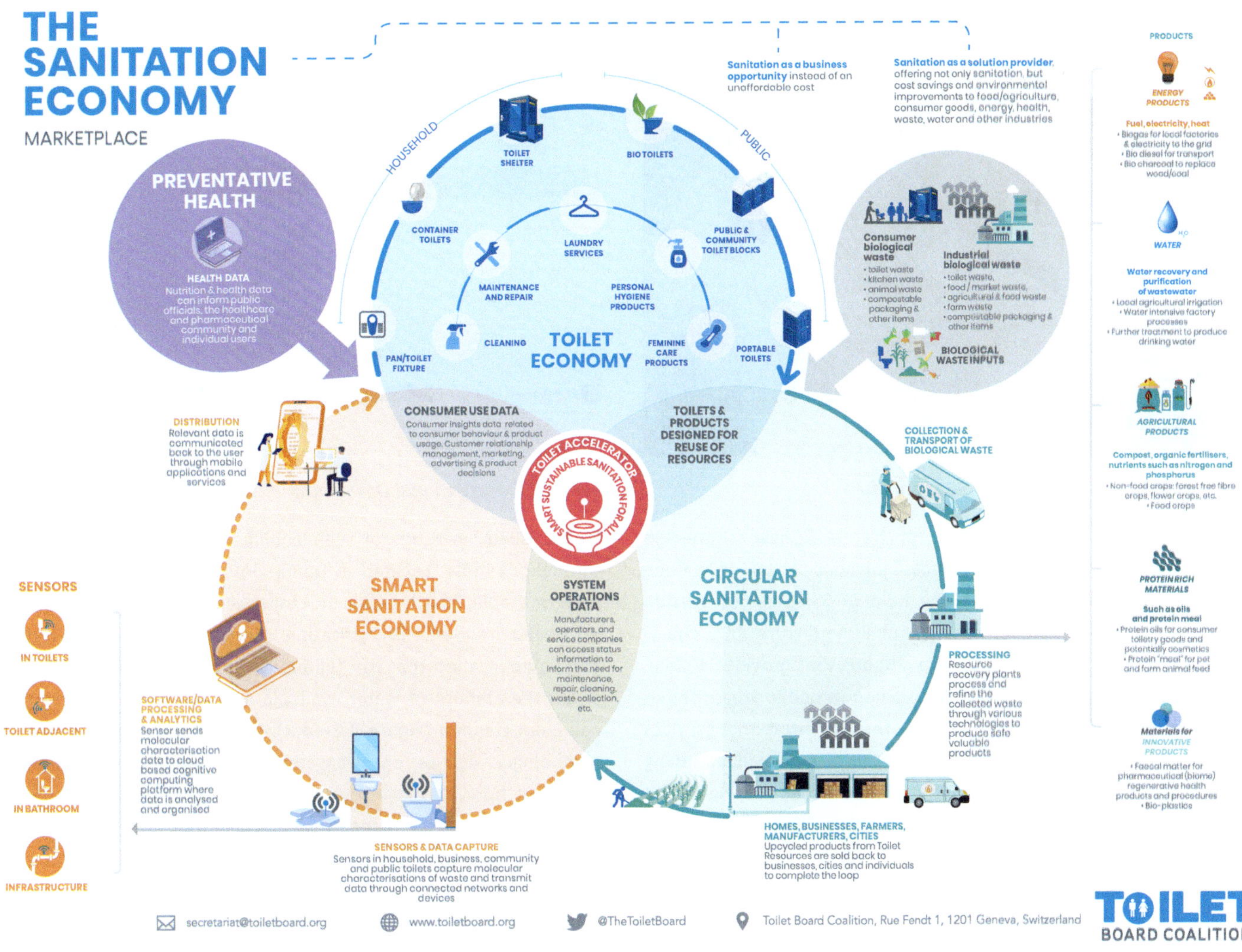

Figure 4.4 Sanitation Economy Infographic, Toilet Board Coalition, 2020

creation. The sanitation intelligence produced from a sanitation economy approach within a city enables more efficient and informed decision-making, leading to vital cost savings for the city, better services for citizens and innovative revenue-generating opportunities in partnership with the private sector (Figure 4.4).

New digital technologies available today – including smart sensors – can be placed throughout sanitation systems in a city to provide real-time user, resource, and health information that will save the city money and create new revenue opportunities for business. Using data captured throughout the system, stakeholders can monitor, manage and respond in real time. In addition, the power of Earth observation, including satellite imagery and geospatial data, can unlock insights that were once invisible. Making sensor and Earth observation data available and transparent enables an open innovation platform for businesses and innovators to provide new solutions that will improve the lives of billions of citizens in cities.

New sanitation intelligence such as increasing use of smart sensors, the power of Earth observation, satellite imagery, and geospatial data unlocks significant cost savings and new revenue potential for cities and businesses in toilets, treatment and health.

Smart cities around the world are applying smart sensor technologies to understand traffic and pollution patterns to improve services. From 2017 to 2019, the city of Pune, India, in partnership with the Toilet Board Coalition, provided an open innovation laboratory in the city to test available digital technologies that could be applied to public toilets and treatment centres. This has led to the testing of new business models and a first attempt to understand the market potential for smart sanitation approaches.

As the world embarks on a massive digital transformation, our ability to capitalise fully on emerging digital technologies and data for sanitation will be one of the leading drivers of sustainable and resilient sanitation systems for the future.

Figure 4.5 Ti Bus Mobile Application, 2020

Toilet integration (Ti): Toilets for Her is a new, connected hygiene centre business model introduced by the Pune Municipal Corporation with the help of Saraplast to provide public toilets for women throughout Pune (Figure 4.5). The Ti bus is a converted out-of-service city bus that provides a clean and safe pay-per-use facility for women in public areas. The Ti Business model generates revenue through additional services at the public toilet, such as a laundry or café, as well as selling goods, such as sanitary pads and products that provide health information. The buses have integrated digital technology to provide an insightful new bases for consumer research, informing potential new product and service information.

Sensors in the buses give information about the footfall, the ventilation and the humidity of the toilet; TV screens display educational communications to promote behavioural change; feedback monitors obtain comments and suggestions about the toilet use; solar panels electrify the buses; and Wi-Fi is used to count the number of people around the bus to understand the frequency and peak hours of use. A mobile application called Soch-O-Mat (meaning don't hesitate) allows the users to locate and navigate to the nearest possible toilets. The Ti toilets aim to extend the use of digital technologies for obtaining data related to the public health of the users, and from the wastewater and toilet resources generated; however, the expansion of such decentralisation will need continued support and political will from Pune city.

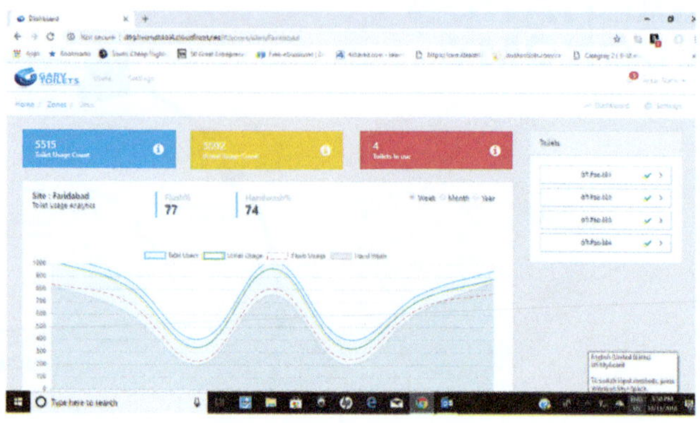

Figure 4.6 GARV Digital Dashboard, 2020

Smart toilets, in terms of self-cleaning, have become popular throughout India. Located in Delhi, GARV Toilets are stainless steel, portable, smart, public bio- toilets equipped with an RFID-IoT (Radio-Frequency IDentification-Internet of Things)-powered mobile application that can tell how many people have used a toilet, the amount of water available and how often the toilet has been cleaned (Figure 4.6—4.8). It is equipped with smart technologies such as sensor-based flushing systems, which clean the toilet floor and lavatory pan, depending on requirements after every use. Digital sensors installed inside the toilet also provide information such as the percentage of users flushing the toilets, the amount of water used, the number of people using soap dispensers, and help monitor the number of male and female users in the toilets.

Figure 4.7 GARV Toilet Interior, 2020
Figure 4.8 GARV Toilet Exterior, 2020

Along with the use of radio-frequency identification technologies, GARV is piloting mobile solutions for pay-for-use, geo-locating and mobile market surveys, as well as advertising on the toilets by the maintenance agency. The radio-frequency identification dashboard is mobile-friendly as well. Through this, the implementing agency will be able to monitor real-time data (health, hygiene, asset related) on any mobile device through the Internet of Things. The new scientific advancements have enabled marked real-time biosensing of wastewater sources.

GARV is currently working on developing biosensors to enhance wastewater monitoring and improve preventative response measures for outbreaks such as COVID-19.

CASE STUDY 7: WISH FOR WASH

Wish for WASH published, in June 2020, a systematic literature review on the use of biosensors in sanitation infrastructure, including toilets, sewage pipes, and septic tanks. The findings from both peer-reviewed and grey literature included sensors measuring a variety of individual and population health outcomes.

At the individual level, these sensors ranged from gathering data about specific health outcomes, such as quantifying drug consumption or screening for infectious diseases, to monitoring individual health-related behaviours, such as tracking latrine and toilet usage. Additionally, various technology types were represented among the results, including DNA biosensors that could detect both cancer markers and pathogens, disposable carbon electrode sensors to detect pathogenic bacteria and smart toilets that quantify CO_2 as a proxy measure for gut microbiome health.

These emerging, Fourth Industrial Revolution technologies also have the potential to gather data about the health conditions and behaviours of communities. For example, "smart" technologies, including biosensors, could more efficiently monitor disease outbreaks in a population.

As one of the papers included in the literature review stated:

> "Infectious diseases require rapid or even real-time detection to assess whether there is a need for the containment of the disease carriers to certain areas and prevent the development of an epidemic. To this end, there is a need to develop novel analytical tools that are able accurately and rapidly to monitor low levels of biomarkers/pathogens with minimal sample processing by unskilled personnel at the site of sample collection."

Additionally, biosensors could be used to supplement standard methods that are already in place, such as colonoscopies, surveys, or blood work testing to monitor individual and population health more holistically and with greater user acceptability.

The literature review showed that biosensors could save time and money compared to traditional data collection methods. There is interest from both academia and the private sectors in prototyping and piloting of biosensors in sanitation infrastructure. The Wish for WASH authors noted that "by

predicting and provoking, sewage systems will become 'smart' and could include 'lab-on-chip biosensors' which would permit continuous data collection and real-time surveillance for viral outbreaks at the aggregate level".

Overall, the Wish for WASH team's findings demonstrate that biosensors are being developed for use in the sanitation sector and could collect useful health-related data. In order to develop and effectively implement these technologies for public health surveillance and individual health monitoring at a large scale, more robust research and product development is needed to evaluate and improve existing biosensor technologies to ensure their efficacy as well as adherence to global data privacy and human study protocols. Additional research in this field could also identify gaps where next generation versions of the Smart Sanitation biosensor technologies can emerge because, as the Wish for WASH team says, #everybodypoops.

A new era for public health – wastewater epidemiology

Environmental surveillance – the process of sampling and analysing environmental samples such as water, air, or surfaces – can provide key information about the presence and dynamics of pathogens, such as polio, typhoid and SARS-CoV-2. Particularly in the framework of wastewater surveillance, WBE refers to the chemical analysis of pollutants and biomarkers in raw wastewater to obtain qualitative and quantitative data on the activity of inhabitants of a given region. WBE has the double role of providing information about the health and habits of the population as well as clarifying the anthropogenic impact on the wastewater. Over the past two decades, WBE has established itself as a valuable tool to track certain societal behaviours and the impact of everyday life on human health.

Currently, WBE is used to monitor the exposure of communities to organic and inorganic pollutants, carcinogens, as well as the abuse of legal substances (caffeine, nicotine, alcohol) and illegal drugs (cannabis, cocaine, amphetamine, opiates, etc.). Additionally, the WBE approach has been used to help identify the presence of diseases (hepatitis, poliovirus, norovirus) and as a tool for testing vulnerable communities. For example, wastewater surveillance has successfully been used as a sentinel system for polio in Israel, triggering a widespread vaccination campaign to prevent further spread.

While the novel coronavirus has destroyed the economic, social and cultural fabric of governance, systems all around the world are rethinking their operation styles and response mechanisms to adapt to the crisis. Through the global response to COVID-19, we have seen a new social contract emerging around health and hygiene. We have seen communities around the world turning to environmental surveillance, specifically WBE, as a method to prepare and respond better to the current and future outbreaks of COVID-19 by monitoring public health at the aggregate level. As a result of the COVID-19 pandemic, the crucial role of WBE in identifying hot spots as well as being a tool for surveillance and early warning in future outbreaks has become evident. The ability to detect silently circulating pathogens, as well as to gain unbiased insights into community-wide pathogen levels and dynamics, can help guide public health interventions such as vaccination campaigns or local shutdowns.

WBE can help detect and estimate the prevalence of pathogens in communities independently of clinical surveillance and therefore without the biases introduced by the availability of clinical testing. WBE is particularly valuable for diseases with long incubation periods and/or high rates of asymptomatic carriers, and it can therefore serve as a population-level early-warning system, as well to evaluate the efficacy of public health interventions in real time. WBE has been used for the early detection and direct mitigation of disease outbreaks in Israel, Egypt and Sweden.

Wastewater surveillance through WBE can provide a vital supplement to clinical surveillance for informing public health decision-making as it offers a relatively inexpensive, efficient means to monitor public health at the aggregate level. WBE can detect and estimate the prevalence of pathogens in communities.

Scientific research has shown the possibility and potential of using these techniques, but their translation from scientific studies to applying the results to advise health authorities is an ongoing challenge. This gap in the current epidemiological landscape points to the lack of a reliable, concrete framework in place through which health-related data can be obtained, processed and transmitted. This lays the groundwork for a potential and promising new ecosystem and data architecture around public health, integrating the data streams available through the sanitation economy.

For sanitation and water operators, this presents a prime moment of opportunity to re-imagine their products and services for hyper-relevance in a post-COVID-19 world. The digitalisation of systems, monitoring and consumer interfaces, and the embracing of developing health sensoring technologies, will generate new and greater value and relevance to already essential services.

COVID-19 is a prime candidate for wastewater surveillance as a complement to clinical testing because it has a relatively long incubation period, high rates of asymptomatic and subclinical infections, and the widespread lack of capacity for adequate clinical testing.

In the global response to COVID-19, several groups have demonstrated that SARS-CoV-2 can be detected in wastewater in the early stages of local outbreaks and that viral dynamics in wastewater mirror COVID-19 cases in those regions

Research and commercial laboratories have now gathered valuable experience in the handling of sewage samples and analysis of the virus.

Wastewater-based monitoring of SARS-CoV-2, both at larger population scales (e.g., at wastewater treatment facilities) and at smaller neighbourhood scales, can help public health officials identify emerging areas of concern and implement appropriate policies to mitigate its further spread, diminishing future strain on healthcare facilities and saving lives. High-resolution (i.e., neighbourhood- or city- level) wastewater surveillance may also enable local governments to deliver "precision public health", introducing interventions only in affected localities and mitigating the economic impact of broad, untargeted policies.

The global response to COVID-19 needs to have the capacity to access the epidemiological surveillance information at national, regional and local levels, through multiple channels, with the ability to monitor real-time data and daily situation reports. To leverage fully the investments and capacities for data collection and analysis for risk assessment, a new global public health data architecture will be required.

Continuing comprehensive and verified global surveillance data about COVID-19 is crucial for response at global, national, and local levels. Epidemiological surveillance information must be collected from all countries, territories and areas, and made accessible through multiple channels, including a dynamic dashboard and a daily situation report, as well as downloadable data extracts.

CASE STUDY 8: MIT/BIOBOT AND COVID-19

As the COVID-19 pandemic began to unfold in the US, Professor Eric J. Alm at the Massachusetts Institute of Technology (MIT) teamed up with Biobot Analytics – a wastewater surveillance company in Cambridge, Massachusetts – to monitor the outbreak. Collecting wastewater samples from the Deer Island wastewater treatment facility, which serves approximately 2.3 million residents in eastern Massachusetts, as well as from upstream sampling points (manholes) representing smaller communities, Professor Alm's research group and Biobot tracked the emergence and rise of SARS-CoV-2 in the greater Boston area. They were able to detect SARS-CoV-2 in Deer Island wastewater as early as 2 March 2020, when there were only two clinically confirmed cases in Massachusetts (Figure 4.9).

Key findings of this work included that SARS-CoV-2 dynamics in wastewater anticipated the dynamics of new clinical cases by an estimated 4–5 days. The strong correspondence between wastewater-based and clinical dynamics allowed researchers from Alm's laboratory to infer individual viral shedding dynamics early in infection, probably before people would have developed major symptoms and sought clinical care. Analysis of upstream catchments revealed strong disparities in viral loads, with increased viral concentrations being found in communities with lower median incomes – consistent with findings that COVID-19 has disproportionately affected socioeconomically disadvantaged communities.

The sensitivity of wastewater-based SARS-CoV-2 surveillance, combined with its anticipation of clinical trends — a critical benefit when dealing with an exponentially spreading disease — emphasises the potential of wastewater surveillance as an early-warning system that is unbiased by limited clinical testing capacity. The ability to use wastewater surveillance to infer information about individual disease course that cannot be obtained in the clinic further highlights the power of this approach. The Alm lab is now extending this work with a COVID-19 surveillance pilot project on the MIT campus that monitors SARS-CoV-2 in the wastewater of individual residential buildings. Sampling occurs inside each building, via sampling ports installed in sewage exit pipes, with results available within 24 hours. Taken together, the Alm Lab and Biobot's work demonstrates the diverse scales at which WBE can be implemented, from populations of tens or hundreds of building residents, all the way up to millions of inhabitants in a region.

CASE STUDY 9: WOODCO

Woodco Bioscience is developing EMPYRE (Energy from Modular Pyrolysis Equipment), a smart and scalable sanitation management solution that combines state of the art organic waste treatment and resource recovery processes with cutting edge internet of things and AI technologies to deliver next generation smart sanitation services.

EMPYRE's core pyrolysis-based waste treatment transforms human waste into valuable thermal and electrical energy and biochar, while also neutralising odours and harmful pathogens. Biochar is a carbon-based product that prevents emissions from biomass that would otherwise naturally degrade to harmful greenhouse gases. It has many applications that are beneficial to human health and the environment. The overall process can be operated off-grid and is a recognised negative emissions technology that can be used to avail of carbon credits.

EMPYRE's disruptive smart sanitation management platform harnesses the latest in internet of things and AI technologies, including space-based technologies (Earth Observation, Satellite Communications and Global Navigation Satellite System (GNSS)), to optimise the waste treatment process and deliver an extensible platform for the development of new applications and services for the sanitation market. The platform uses a combination of terrestrial and space-based sensing capabilities to collect contextually complementary data sets relating to sanitation management and operational processes, toilet-user behaviour and health, sanitation-related pathogenic activity, sanitation-related impacts on water quality and broader environmental factors, such as weather and ground conditions. User inputs from stakeholders within the sanitation ecosystem are linked through web/phone apps. Combined with various cloud-based services, powerful analytics and intuitive user-interfaces, EMPYRE can be used to track, monitor, detect, react to, and even predict sanitation system behaviours and outcomes.

A key goal of the initial deployment of EMPYRE is to provide predictive health analytics through the identification and monitoring of risk factors associated with outbreaks of sanitation-related disease and to provide an early warning system. Examples of bespoke AI-driven sensing capabilities being developed under the project to support this goal include real-time pathogen detection in wastewater and waterways and diarrhoea detection in community toilets. Earth Observation will be used to monitor the impact of illegal dumping of faecal sludge in watercourses, and to track the flow of waste and its interaction with the local environment and local populations. Woodco Bioscience envisages that such sensing capabilities will play a critical role in predicting new outbreaks of disease and in tracking the evolution of outbreaks as part of its smart sanitation management solution.

EMPYRE strives to deliver next generation sanitation services for a green and circular sanitation economy. Its extensible "single pane of glass" view of the sanitation ecosystem will provide a range of new actionable insights not readily available from any single data source. EMPYRE's off-grid and satcom capabilities mean that it can be deployed even in remote locations where there is limited existing infrastructure. The solution will be deployed for demonstration in Durban, South Africa, in Q3 2022, supported by the European Space Agency (ESA) following a successful "Space for Sanitation" feasibility study.

4.2 Case studies on the use of digital tools:
The use of digital transformation in the COVID-19 pandemic

The COVID-19 pandemic has been an unprecedented challenge for local and health authorities since its onset in December 2019, both because of the severity of its impact on the population and the economy, and because of how long it has lasted.

Although there have been different approaches in the fight against the virus during the 2 years of the pandemic, there has always been a common element: the lack of information for swift, effective decision-making.

PCR testing has definitely been the most common tool used by countries to identify positive cases in the population. However, the effectiveness of these tests relies on their massive, repeated use to prevent uncontrolled outbreaks, and many countries were faced with the difficulty of locating asymptomatic cases, which caused the virus to spread silently among the population.

Against this background, WBE provides the authorities with very useful information. By establishing the concentration of genomic traces of the SARS-CoV-2 virus in wastewater in the laboratory, its incidence could be linked to the sector of the sanitation network from which the sample had been taken. More importantly, this genetic concentration was 7—10 days ahead of the epidemiological reality that would later appear in the population in terms of contagion. Therefore, a single PCR could be used to analyse all the inflows to the wastewater network from the first day of infection, including both symptomatic and asymptomatic individuals who could spread the infection but had not been detected by the health service (see Figure 4.9).

This situation made wastewater utilities the main allies of the health authorities who began to request WBE-related data.

However, for this WBE data to become an effective early warning system, a digital tool with the capacity to integrate and unify all the WBE-related data sources needed to bring all of the different data sources together in order to display, in an easy to use format, so that the insights from the data could be easily realised.

The different data sources added, included the following:

• Geographic Information Systems (GIS) of the sanitation systems to identify the basins in the network that collect the wastewater from the target population sectors.

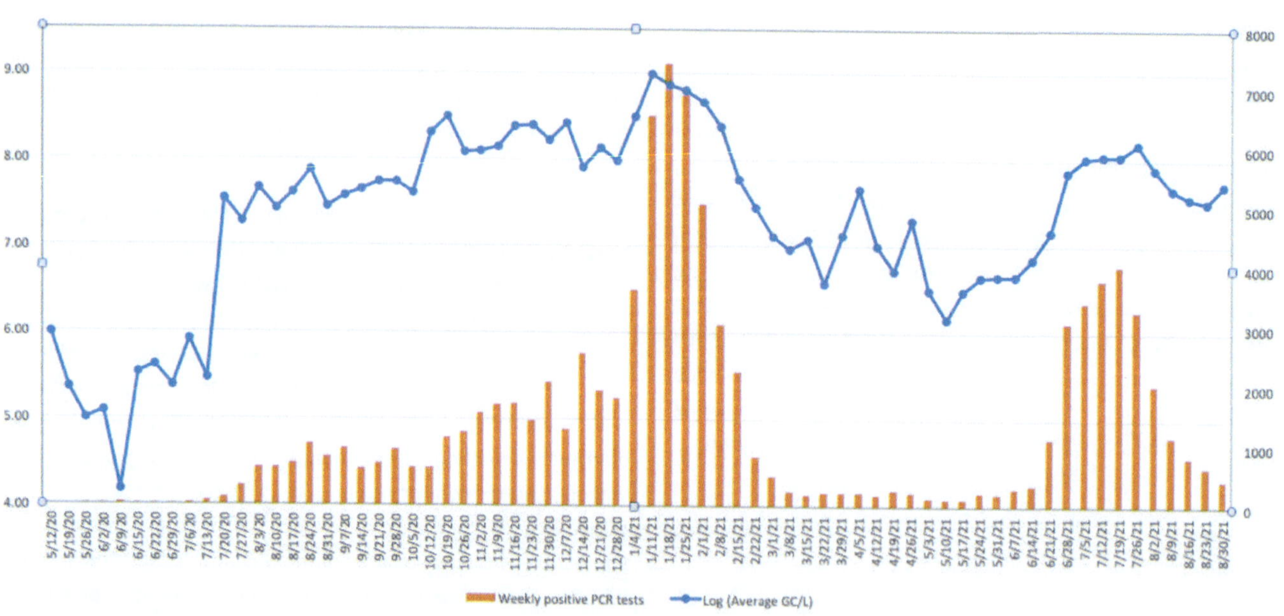

Figure 4.9 Weekly evolution of positive PCRs and concentration of SARS-CoV-2 genomic traces in wastewater (source: Global Omnium)

- Cartographic system for the geolocation of the samples analysed and of assets and special places of interest such as hospitals, health centres, elderly care homes, and schools.

- Laboratory Information Management Systems (LIMS) for rapid collection and display of analytical results.

- Third-party applications, such as official population statistics data and hospital occupancy rates.

- Water consumption from smart meters to standardise the virus concentrations obtained.

The GoAigua SARS Analytics software was specifically designed by Idrica to provide a solution to this challenge by offering water utilities a vital tool to set up an early warning system based on WBE.

Since its launch, GoAigua SARS Analytics has been successfully deployed by the utility Global Omnium in various environments in Spain, assisting local governments and health authorities in protecting public health and the economy.

The following are some of the utility's success stories in managing the water cycle in more than 400 cities.

USE CASE 1: PROTECTING VULNERABLE POPULATIONS

As in other neighboring countries, the SARS-CoV-2 pandemic in Spain was characterised by a rapid spread of the virus and an unequal impact among the different population age groups. Statistical data published by the Spanish Ministry of Health in May 2020 shows an explosive increase in COVID-19 mortality in the over 60 age groups (see Figures 4.10 and 4.13).

The mortality rate recorded at the beginning of the virus outbreak exceeded all forecasts, putting the health services under unprecedented levels of stress and strain (Figure 4.11).

In this context, protecting the population over 60 years of age became the main priority for local authorities in Spain. However, as in other neighboring countries, the Spanish healthcare system was overwhelmed at the time, making it unfeasible to carry out mass PCR testing of the entire population.

The use of WBE by water utilities provided the relevant authorities with the information they lacked: where and when to carry out PCR tests well in advance in order to prevent deaths.

Global Omnium was commissioned to conduct repeated testing of wastewater from all the elderly care homes in one of Spain's major regions to rapidly detect the existence of the virus and its evolution so as to prioritise PCR testing by health authorities, pre-empt new outbreaks and even determine the timing of vaccination campaigns.

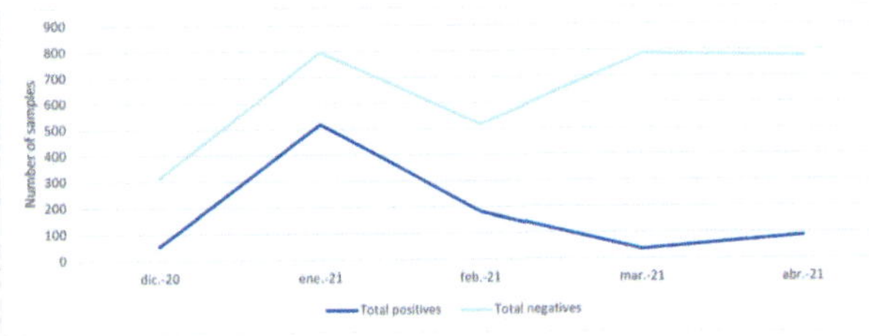

Figure 4.10 COVID-19 mortality by age group (source: Ministry of Science and Innovation, Ministry of Health, National Health System Interterritorial Council and Carlos III Health Institute. Report on the situation of COVID-19 in Spain).

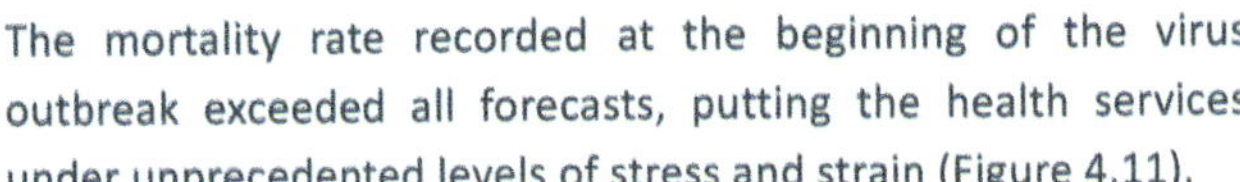

Figure 4.11 Evolution of daily mortality (source: Ministry of Science, Social Services and Equality, Spain).

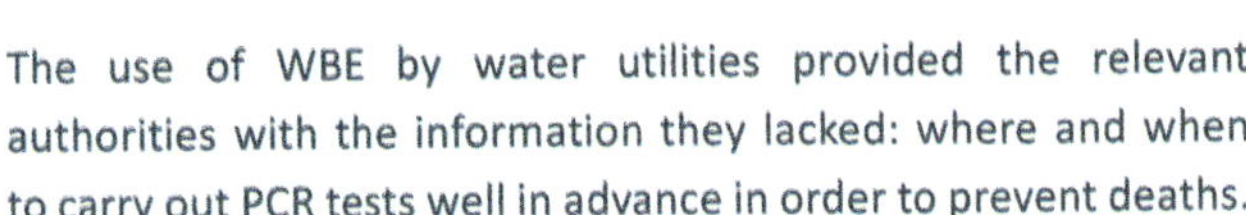

Figure 4.12 Monthly evolution of the detection of SARS-CoV-2 in wastewater from elderly care homes (source: Global Omnium).

Work was conducted between December 2020 and April 2021, with weekly analysis of the wastewater from care homes. A total of 4,066 SARS-CoV-2 tests were carried out, of which 875 were positive (see Figure 4.12).

Thanks to the weekly monitoring of wastewater results, the relevant authorities were alerted about the spread of the virus in the region's elderly care homes, thus serving as an early warning system to optimise both lockdowns in homes and subsequent PCR testing. The graphic in Figure 4.13 shows the positive weekly evolution in care homes.

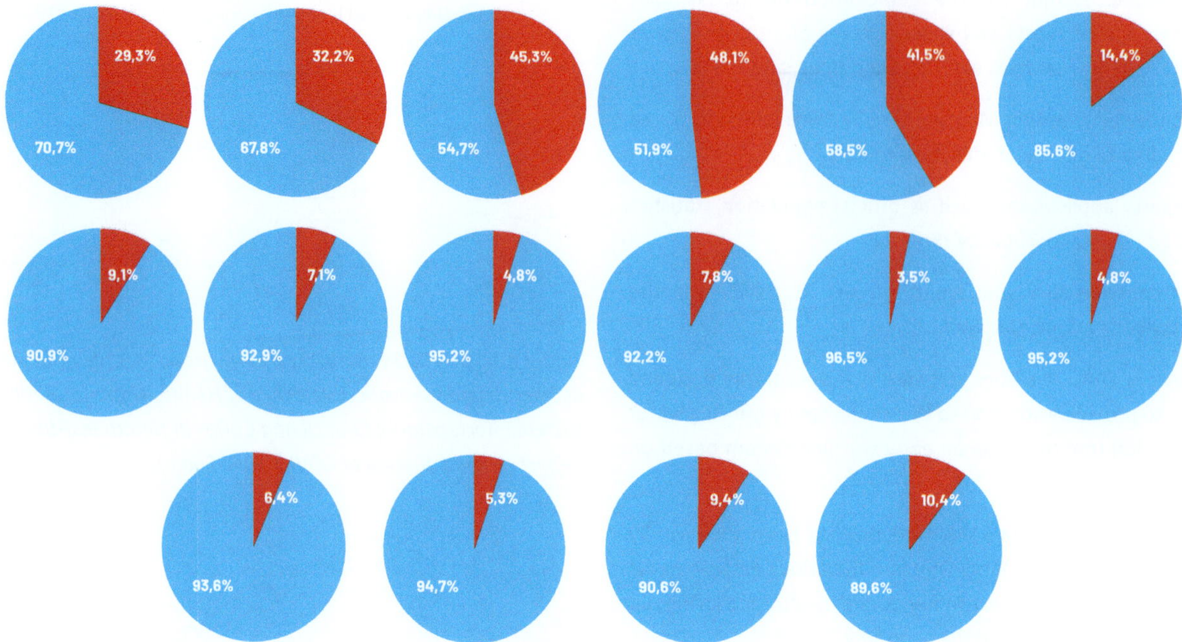

Figure 4.13 Weekly evolution of the detection of genomic traces of SARS-CoV-2 in wastewater from elderly care homes. In red: positive cases. In blue: negative cases (source: Global Omnium).

Detecting the virus in wastewater meant that the authorities could react quickly and promptly to isolate positive cases, thus preventing uncontrolled outbreaks and saving lives among the most vulnerable groups of the population.

During the course of the work, the GoAigua SARS Analytics software was a fundamental tool that provided the utility Global Omnium and the health authorities with real-time information on analytical results and their geolocation. In addition, the software offered useful data embedded in the tool, such as the number of residents in the homes.

This minimised the number of PCR tests to be performed and optimised the number of healthcare staff needed to carry them out.

USE CASE 2: PROTECTING THE ECONOMY

Throughout the pandemic, population lockdowns and restrictions on international travel have been ongoing. Several countries in Europe decided to restrict travel to other European regions to a greater or lesser extent depending on the reported SARS-CoV-2 incidence. This situation had a major economic impact on the main tourist regions in Spain, whose GDP depends to a large extent on this industry.

Therefore, it became imperative for tourist areas to make quick and effective decisions in order to keep the cumulative rate of the virus under control and thus maintain their region on the European list of areas eligible to welcome international travelers.

In this context, the SARS-CoV-2 early warning service offered by the utility Global Omnium and supported by the GoAigua SARS Analytics tool was of vital importance on the Spanish tourist island of Lanzarote (Canary Islands) in attaining this objective.

Detecting asymptomatic cases early on became the main headache for the health authorities, as these cases were outside the regular loop of ongoing PCR testing and posed a serious risk of a viral outbreak with major social, economic and health consequences. For this reason, the authorities decided to entrust the utility Global Omnium with the monitoring of the entire island by sectors and specific points (see Figure 4.14).

Geolocating the concentration of the SARS-Cov-2 virus in wastewater enabled the authorities to compare this on a weekly basis with the PCR tests performed in order to improve the weekly planning of these tests. Thanks to the correlation of the virus concentration in wastewater with its source, major information campaigns were organised to inform the population by offering PCR tests in neighborhoods with a high probability of asymptomatic cases.

The GoAigua SARS Analytics tool provided all the information from the catchment basin at the monitoring point, which was a key element in detecting the neighborhoods and houses where testing needed to be offered. The added value provided by WBE led to the development of GoAigua BioRisk, a technological solution for monitoring not only the presence of viruses, such as SARS-CoV-2, but also of other substances in the sewer network.

The use of digital transformation in sewer blockage detection

The use of Digital Transformation in the COVID-19 global pandemic is only one aspect of how Digital Transformation can assist the Global Water Industry. In this case study, we look at one of the digital applications that has been increasingly developed in recent decades, sewer blockages. The misuse of sewers, in the UK alone, costs £200 million/annum. This is mainly through the use of wet wipes and the treatment of the toilet as a "wet bin". This is causing a large number of pollution events being detected through increased monitoring of the sewer network. It's not that the problem didn't exist before but it is the increased detection that is highlighting the problem. In 2021, the UK saw 2.67 million hours of spills to the environment which was a decrease from the 3.1 million hours the previous year.

This is where the supply chain has been taking on the technological challenge and using digital tools and data from various sources to trigger an alert system of where potential blockages are developing in the wastewater collection network. This enables water companies to specifically target areas of blockage and minimise the pollution events and consequently the impact on the environment.

One such technology was trialled by the English water company, Wessex Water, in the City of Bath in the South-West of England. Sewer blockages were a challenge to the water company which

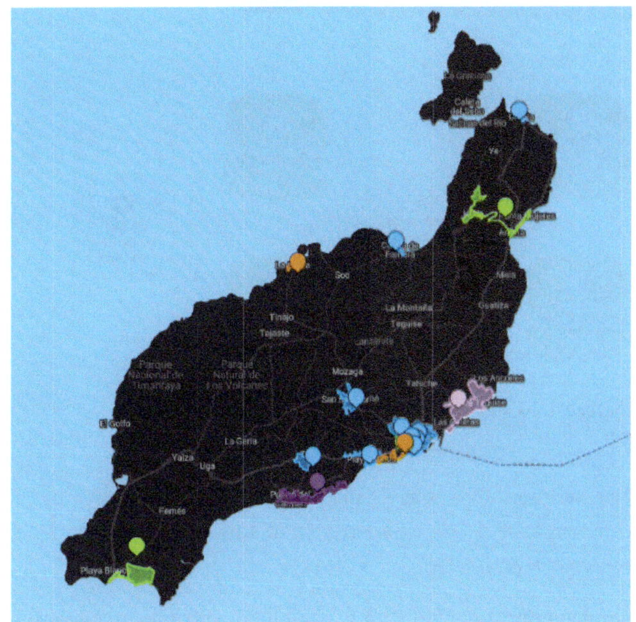

Figure 4.14 Geolocation of wastewater monitoring points on Lanzarote using the GoAigua SARS Analytics tool (source: Global Omnium).

has 35,000 km of sewer and serve 2.8 million customers. They challenged the supply chain market as shown in Figure 4.15.

The challenge was resolved by using a blockage prediction system that used telemetry data. This included the sewer level from Wessex Water's telemetry system along with hyperlocal rainfall data which predicts rainfall over 1 km grids. (which enables a localised prediction of rainfall and the impacts that it should have). This can then be compared to what it is actually doing by using the level data in the sewer network.

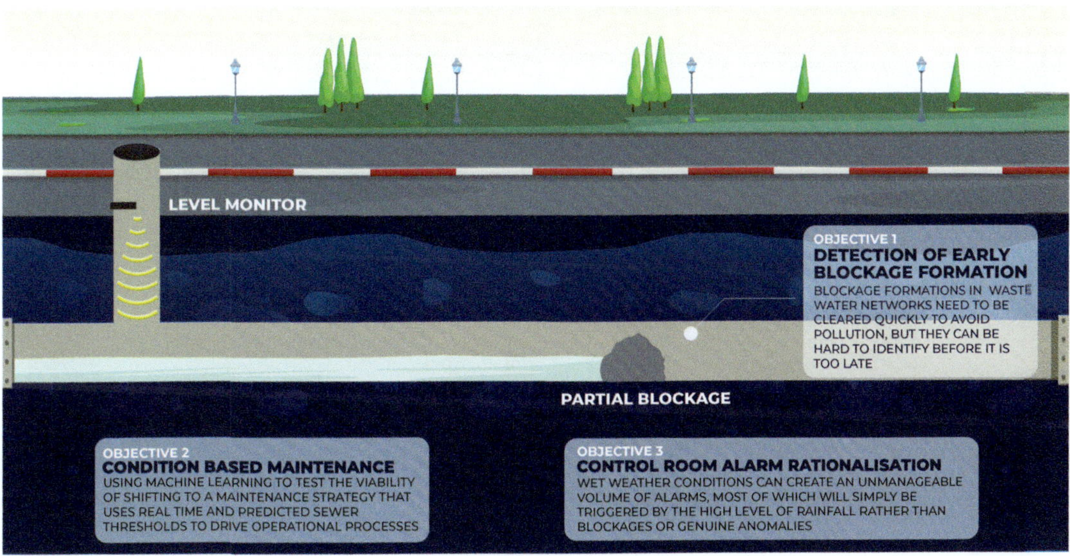

Figure 4.15 The blockage challenge as set by Wessex Water

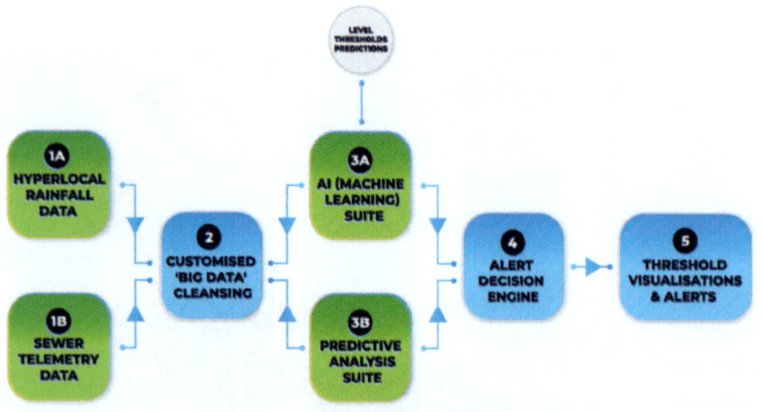

Figure 4.16 StormHarvester blockage prediction.

The solution that was trialled in Wessex Water was by a company called StormHarvester and they used a process as shown in Figure 4.16.

The hyperlocal rainfall data is combined with sewer telemetry data from both CSOs and pumping stations. The combined data undergoes data cleansing before being processed in StormHarvester's proprietary machine learning system which uses both algorithms and predictive analysis tools in order to predict what the levels of wastewater in the network should be in order to determine where potential blockages are forming in near-real time.

Figure 4.17 shows some of the results: we see the actual sewer level as measured by a sewer level monitor (brown) compared to the predicted sewer level in the bands around the actual sewer level. In Figure 4.18 we see rainfall happen within the catchment and the sewer respond and spill to the environment as the rainfall increases. If there is a situation where the prediction differs from the actual levels as measured by the sewer level monitor, this would trigger an alert in the control room for operatives to investigate a potential blockage.

In this case study, the measured level exceeds the predicted level due to an actual blockage within the sewer caused by settled debris in dry conditions. The level increases to above what the system considers normal and an alert is sent out. A jetting crew is sent out to relieve the situation and clean the sewer. This is ineffective and CCTV is used to diagnose the problem and enable a repair crew to fix the problem within the sewer.

The key considerations within this case study are:

- The blockage was predicted in dry weather conditions which highlights the sensitivity of the system and minimises the risk of a pollution event every happening because of this sensitivity.

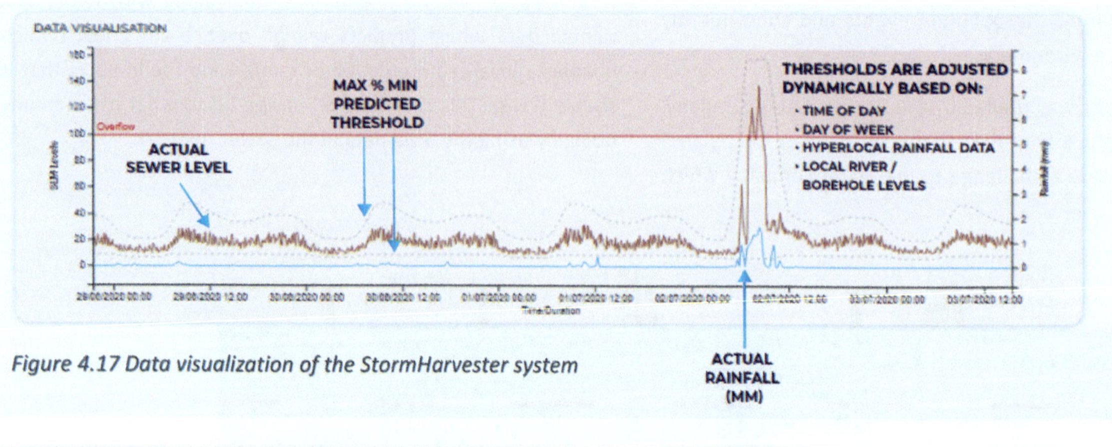

Figure 4.17 Data visualization of the StormHarvester system

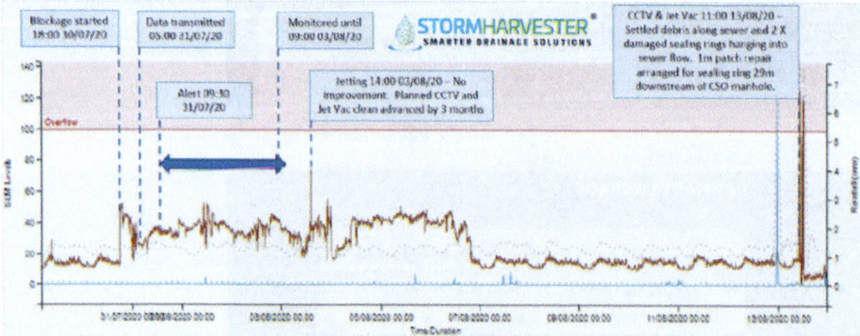

Figure 4.18 A blockage time of a detected blockage

- There is a delay between the sewer becoming blocked and an alert happening and then being reacted to. This highlights the importance of getting any predictions correct and minimising the impact on resource.

- When jetting was unsuccessful, the system indicated that a problem still existed. This allowed Wessex Water to take further action and escalate the maintenance work that was necessary to resolve the problem whilst minimising any impact on the environment as the issue happened in dry weather.

The system was trialed for a period of 3 months within the single catchment which served the City of Bath and 60 early blockage formations were detected and two significant pollution events averted. The other benefits of the system were that it minimises false alarms but also as the partial blockages are detected early they have less time to worsen. This is especially a problem with fats, oils, and greases which can set like concrete and become difficult and costly to remove. The system had a 92% alert accuracy in early blockage formation prediction and sensor anomaly identification and reduced the control room alerts by 97% and did not miss any unpermitted spillages due to blockages within the network.

In this case study which was originally a proof of concept trial which has now become more widely adopted within Wessex Water, we see the benefits that Digital Transformation can bring to the water industry as a whole. Blockages across the sewer network are costly and put customers and the environment at risk. In this case study it was sealing rings and debris; in another case study, it was an ordinary house brick. In other scenarios, it will be wet wipes or fats, oils, and greases causing the problem. Amongst the vast number of alarms that are raised in a water control room, it is very difficult to pick out what data is giving a true reflection of a situation and what is caused by poor weather conditions, bad sensor data, or a whole host of reasons. In the blockage prediction case study, the key is bringing the data together and presenting it as an event that needs to be managed focusing the control room operative and ensuring that the operational teams in the field are used most efficiently to resolve the issues. This is situational awareness, which has to be one of the outputs from Digital Transformation, at its best.

Conclusion
Application, application, application!

In this chapter, we have obtained an overview of what Digital Transformation can do for our society and have considered some important applications in water, sanitation & hygiene (WASH) and how digitalisation can help to achieve SDG6. When applied in response to the SARS-COV2 global pandemic, digital tools can limit the spread of the disease or even help the world detect future pandemics. Finally, we have described how digitalisation can act as a tool to help us manage modern life and help prevent pollution events.

However, there are many other areas of applications for Digital Transformation that the global industry is developing right now to address issues such as climate change and to reach Net Zero. These current challenges and the future of digital transformation will be discussed in the next chapter.

Chapter 5

Digital transformation — now and moving forward

Introduction

What is the future of digital transformation? What do we need to do to transform? There is a large number of case studies which investigate the adoption of Digital Transformation, from organisations that are at the front line of water and sanitation to more traditional water utilities providing water services to millions of customers. It is being delivered in pockets of good practice and there are barriers to the adoption of the approach whether technological, financial or psychological. In this chapter, we look at what the "Digital Future" is and what are the barriers to its adoption from the perspective of experience practitioners as well as young water professionals (YWP) in IWA.

A digital future in the here and now

The challenges faced by the global water industry are numerous:

- In 2020, only 74% of the world had access to safe clean water (WHO, 2022).

- In 2022, the world accepted that a preventing global temperature rise of 1.5°C wasn't really achievable and, as a result, the impact of climate change is only going to increase. This will increase sea levels.

- Water stress and water resources issues across the globe are aggravated by global water leakage which is estimated to be in the region of 45 million m³/day (World Bank, 2016).

- The "Race to Net Zero", the global challenge to help to mitigate climate change, targets 2050; however, practitioners hoping to complete the challenge must act much more rapidly than this.

There are also more regional issues. In fact, in the UK, there is a water pollution crisis with not a single water course achieving good chemical status and only 45% achieving good status for phosphorus pollution through a combination of agricultural and industrial pollution (EAC, 2022). Although these issues have been highlighted in the UK, the likelihood is that the issues also exist elsewhere.

These come across as almost unassailable challenges for the global water industry, however, this is where Digital Transformation can and must be used. Nevertheless, there are some steps to take and some challenges to face.

The challenge of data

The global water industry collects millions of pieces of data every day. With the advent of digital transformation, the amount of data is destined to increase even more.

We have seen this with smart water meters in people's homes where utilities have moved from twice annual meter reads to hourly data, allowing customers to pay only what they consume. As a result, customer's consumption has dropped. This is because water is no longer seen as an unlimited resource for somebody who is paying a fixed rate. Water meter data helps in company leak detection and planning. However, it comes at a cost and IT systems have to be updated and strengthened to cope with the amount of data that has to be handled.

Nevertheless, increasing the amount of data has a limit and raises some concerns. For example, by reading water consumption of a house every hour, it could be possible to detect when or if someone is actually in the house. With a minute-by-minute water meter, it could be possible to tell when a house is running a washing machine or somebody is taking a shower. This could understandably cause people to have concerns about invasion of privacy. This is why this sort of data is heavily protected and cybersecurity within the water industry is a challenge that is addressed very strongly and also why there has, traditionally, been a nervousness around the adoption of Digital Transformation, at least in some areas.

Taking a step back into data, it is crucial for this to have an infrastructure. The water industry has always been labelled as "Data Rich, Information Poor". In reality, the management of data and its quality are the first challenges of Digital Transformation. It is something that was described heavily in Chapter 2, as without good quality and accurate data, the Digital Transformation of the water industry is doomed to fail.

The way to overcome this challenge is for the data to have value by relating it to informational need that has purpose. If the value is in place, then the investment will be in place to operate and maintain the data effectively. This includes having a data structure, by the use of meta-data.

This leads to the next challenge – cybersecurity.

The challenge of cybersecurity

Cybersecurity is a challenge that all industries face. The water industry, being part of the national critical infrastructure and holding sensitive data of each of its customers, is not an exception. As such, there are limitations as to what can be done on the grounds of cybersecurity. The control systems of the water and wastewater systems cannot and should not ever be connected to the internet as this is a major security issue. Equally, the protection of customer data is fundamental as a breach of this data has huge implications on the trust of the customer and of course for the resultant financial implications that may bring.

There are challenges associated with the data that the water industry collects and transmits. Firstly, the data has to be categorised in terms of its risk and implications. For instance, flow and quality data from a wastewater site is relatively low in its security implications and the transmission of this data is considered low risk. However, when transmitted back to a site, the implications are higher as there is a possibility to affect the control of the wastewater treatment work. This becomes a very different risk to the operational processes and, as such, it is considered a cybersecurity risk.

This risk, which works at physical level, involves the treatment and network systems and, as these become more digitally transformed, the cybersecurity risk becomes greater and greater. As such, more management and mitigation have to be in place before the advances that Digital Transformation can bring are put into place.

The challenge of legacy systems

Quite often, the biggest response as to why we haven't already "Digitally Transformed" is that the water industry has legacy assets that are not yet ready to be changed. Therefore, until those are not changed, any new implementation has to able to work with existing assets. We simply cannot tear down the whole system and rebuild it from scratch: the cost is simply too much. Additionally, the capacity of the system has its limitations.

This has been seen when smart water meters have been implemented on a large scale. In these occasions, the IT systems and the systems that store all of the meter readings (the data gatherers) have had to be majorly upgraded and separate systems had to be put in place, which is undesirable as it creates a data silo. This problem exists across the entire industry as, once installed, the asset life of equipment tends to be from a minimum of 7 years (some water quality instrumentation) to between 15 and 20 years or even more. Quite simply, the legacy systems cannot cope with some of the major changes that innovation and Digital Transformation can bring.

There is an example of this in the instrumentation used in the water industry. Most modern instrumentation has a wealth of data that will indicate its operating state and how it is performing on top of the critical piece of data – what it's actually measuring. This additional data shows if there is a malfunction. An operator can diagnose the problem remotely and know who to contact to fix the fault. However, the way the industry transmits the data from the site restricts the information that can be sent. Manufacturers have worked around this by enabling remote transmission of the information but cybersecurity, at the moment, is a barrier for these systems to be implemented.

The opportunities that Digital Transformation brings

The Digital Transformation of the water industry brings a number of opportunities that can address current challenges both at a regional and global scale. There is a huge potential for a large transformation.

How precisely we can achieve this Digital Transformation is going to rely on how much it costs and what value it brings to the water industry. At the centre of Digital Transformation is how we, as an industry, use our data and if that data is of a sufficient quality to use.

We've only touched on what the drivers and tools are for the Digital transformation.

The next sections present a collective idea of the future of Digital Water, gathered by the authors from a globally diverse group of Young Water Professionals (YWP), across IWA's water-wise principle levels.

Figure 5.1 The "Principles for Water-Wise Cities" Framework: four Levels of Action – 1. Regenerative Water Services; 2. Water Sensitive Urban Design; 3. Basin Connected Cities; 4. Water-Wise Communities. Source: IWA Water-Wise Cities Principles

5.1 A Digital Future from the future's perspective

Digital Water. Smart Water. Internet of Water. Water 4.0. These are some of the many terms that are being used to describe the rapid transformation that the water sector is experiencing (Sarni et al., 2018). Digital technologies will be integral to addressing the most pressing challenges of our time, namely climate change, population growth, public health crises and the provision of basic services at risk from global shocks. To successfully address these challenges, it will be crucial to embrace a multi-stakeholder approach to problem-solving as well as benefit from all relevant experience and expertise. Additionally, a redesign of education and training will be needed to meet the needs of the water sector of the future while adopting an agile approach. The digital water transition has to aim for open data and data standardisation. In fact, open sharing of lessons learned and solutions ensures progress is not limited to countries with the requisite resources. By doing so, the digital transformation of the water sector should epitomise a 'no one left behind' approach to ensure ongoing equitable and sustainable development. For this purpose, innovation- enabling environments need to be the norm, including innovation-enabling policies, financing, risk tolerance and more. Digital technology must be focused on needs and adopted at all levels – described in the IWA Water-Wise Cities Principles as regenerative water services, water sensitive urban design, basin connected cities and water-wise communities (Figure 5.1).

The water sector needs to create future proof enabling frameworks that support safe, efficient and effective technologies. As a sector where failure is not an option, we cannot apply the technology sector's mantra of "move fast and break things"; rather we need to "move fast and fix things".

Changing the game

As we look ahead to the challenges of the next 5, 20 and 50 years, it is evident that there needs to be a revolution not only of the water sector's technology, but also of its policies, governance and culture. Most importantly, the digital transformation of the water sector should drive sustainable development and deliver a more equitable (water) world.

Water-Wise basins

Globally, basins are under pressure due to floods, droughts, water quality issues (e.g., intensifying toxic algal blooms) and reduced water availability because of competing pressures on resources. It is estimated that around 48% of the world's wastewater flows back into our watersheds untreated (Jones et al., 2021) with drastic socioeconomic, health and environmental consequences. According to IWA's Action Agenda for Basin-Connected Cities, *"protecting basins and restoring those that are already degraded is a priority to ensure a balanced approach to development that sustains cities and the ecosystems they rely on"*. Well-managed basins can be the difference between water security and water scarcity for urban and rural stakeholders. Managing the top three risks identified by the Action Agenda: "extreme events, declining water quality, and water availability" requires an integrated and data-driven approach. Added to these risks is the sheer cost of maintaining infrastructure to buffer the impacts of extreme events and manage water quality and availability. Integrating nature-based solutions into long-term planning supports the triple bottom line but requires data-driven development and decision-making. Digital water tools can support water professionals and basin level stakeholders with the pathways to action from assessment, planning and implementation.

We envision a future in which Integrated Water Resources Management (IWRM) is mainstreamed, urban water is holistically managed, and where nature-based solutions complement grey infrastructure. To help transform this view into a reality, digital tools have a central role to play. For instance, watershed Digital Twins, underpinned by solid real-time data from internet of things sensors and remote sensing, could be applied to support water management decision-making processes. Additionally, Artificial Intelligence (AI) could be used to monitor, evaluate and analyse flows and processes at multiple scales within the basin. Also, predictive modelling capabilities would improve transparency and facilitate knowledge sharing between upstream and downstream stakeholders. The combination of these and other technologies will optimise the potential for basins to provide high quality ecosystem services to the world's cities and rural areas, while maintaining regenerative ecosystem flows.

Pilot programmes and proof of concepts are laying the foundation of what a digital future will hold at the basin level. On the data management and governance level, initiatives to standardise data and create open data platforms are transforming how basins are managed and leading to accelerated innovation. Various projects integrating digital tools (e.g., blockchain, new imaging technologies coupled with augmented reality (AR), drones, etc.) are proving the value of advanced technologies for improving water management at the basin level. However, some technologies are more mature than others. For example, the use of remote sensing, satellite imagery and advanced imaging technology is well documented, and these technologies are already well embedded across the water sector. Where mature technologies meet less mature technologies, things get exciting. For instance, algal blooms in surface water due to agricultural runoff is an increasing problem of global proportion, causing environmental and economic damage to regions (Ho et al., 2019). Stakeholders can better manage the risks and impacts by using Machine Learning (ML) to predict when, where and for how long these blooms will happen (Yñiguez and Ottong, 2020). Advanced Earth-Observation data products are currently being used in the framework of the European H202 project PrimeWater which generates information on the effects up upstream changes on both water quality and quantity. For example, in the Mulargia reservoir (Italy), PrimeWater is assessing the predictability of algae blooms for time scales spanning from several days to a few weeks ahead employing both process-based and data-driven models.

Game-changing technologies like digital twins, internet of things, blockchain, remote sensing, Fintech and AI can allow for information-driven decision-making, planning and financing to a level we have not yet implemented basin-wide. Early investigations by groups in Denmark, the Netherlands and the UK are exploring the possibility of creating a national-level digital twin of their basins to improve how water is managed and used from source to sea (Bolton et al., 2018). Integrating AI for scenario planning can support stakeholders in data-driven decision-making. In cases where trust and transparency are problematic, blockchain can be used in water markets to bypass traditional ways of trading water rights and implement traceable and transparent foundations for water allocations.

Some of the benefits of these technologies being used at scale include real-time and predictive insights, preventative management capabilities, transparent and integrated information sharing, as well as accurate water flow management.

Successful pilot projects will accelerate scaling and adoption, although full-scale adoption will take time and coordination efforts. The data challenges are not insurmountable but will require a coordinated approach among all stakeholders and the necessary investments to ensure that quantity and quality are at the level needed to move towards actionable intelligence. Innovative financing mechanisms that support digital basin scale projects will help establish more of these successful proof of concept projects.

Water-sensitive cities

In the face of increasing pressure from climate change and urbanisation, many cities around the world are struggling to provide basic utility services, such as clean reliable drinking water or sanitation. There is a need for investment in the most basic of infrastructure. Digital tools typically cost a fraction of the investment in 'hard infrastructure' but have the potential to result in significant savings by optimising existing infrastructure or delaying the need for developing new ones. For example, water loss is a significant issue in many cities of the world, where up to 60% (46 billion litres) of clean drinking water is lost on its way to users (Serranito and Donnelly, 2015). This is an issue where digital tools have proven highly effective. For example, the City of Cape Town has identified the need for a Decision Support System (DSS) to support its response to a recent severe drought. This system is current being implemented (Kamish and Cantrell, 2018). Another successful case is The Danish LeakMan project, which is applying digital solutions to reduce urban water loss in greater Copenhagen, Denmark, by combining multiple data sources in automatic leakage management. Noise loggers for automatic leakage tracking optimise leakage detection; smart water meters measure end user consumption and intelligent valves and pumps enable active pressure management. All data collected is combined with SCADA and GIS in real-time hydraulic modelling and processed in online management information systems to facilitate automatic leakage management and monitoring of KPIs, resulting in a significant reduction in Non-Revenue Water levels (Mikkelsen, 2018).

Cities have started utilising digital tools for many applications such as modelling flood risk and forecasting drinking water demand (Nkwunonwo et al., 2020). This trend will only strengthen in the years to come, but will require large investments in digital capabilities. The 'Smart City' market is expected to reach 252 billion USD by 2025 (Smart Cities Market Size | Industry Analysis, Growth, Trends, Forecast 2025), with further growth expected beyond this.

The water sector must be an integrated part of the Smart City of the future. Ideally, water should be linked in autonomous digitally driven systems to the rest of the functions of the city which utilise real-time data and digital twins to ensure holistic and optimised management of all the services and resources. It will no longer be enough to optimise within our own sector: the water sector needs to prepare to 'plug in' to a larger platform that enables the integration of all utility services provided in the city.

A Smart City minimises its environmental impact through reuse and minimal resource consumption, in the spirit of sustainability. At the same time, cities must be as liveable as possible and provide safe and reliable water and sanitation services equally for all. In this framework, digital tools can improve the resilience of cities to water scarcity or flooding by optimal water allocation and forecasting (Groppo et al., 2019).

Moving from today's cities towards integrated smart cities will require investment in hardware, software, processes, and most importantly, people. Cities need to begin investing in distributed networks of sensors to improve a better management of valuable resources (i.e., monitoring, control, reuse), as well as reduce the impact of the water system on the quality of life and environment in the city. This will ensure that new infrastructure to supply, drain and treat water can be optimally designed, maintained and operated through the use of digital tools (Lund et al., 2019).

Digital tools help optimise the linkages between smart city systems. Data is still treated in silos within organisations and between separate utilities, leading to a suboptimal use of the resulting information. Attempts to create uniform data platforms are appearing and will have to be deployed at scale for cities globally (European Commission, 2020). For example, the European Union is investing in creating a digital marketplace for smart city solutions that will bring together key sectors. This cross-sectoral collaboration will require water professionals to acquire an understanding of interoperability as a core design component of digital systems.

For cities to become Smart Water Cities, there is an urgent need to ensure that water professionals, including YWPs, develop the necessary digital skills and integrate work plans across sectors. This requires capacity development on digital skills, systems thinking and working in multi-disciplinary teams with experts from other areas of the Smart City Network.

Water-Wise industry and utilities

The world's water infrastructure needs to keep pace with the increased growth in demand. Assuming business-as-usual, the global demand for water in the manufacturing industry alone will grow by 400% between 2000 and 2050, mainly from emerging economies (WWAP, 2015). Additionally, existing water infrastructure is ageing and faces large scale investment needs and associated control and design decisions.

The negative environmental impact of industry and ill-managed water systems is contributing to global water pollution. It is estimated that less than 20% of the world's drainage basins show near pristine water quality (WWAP, 2009). In this context, digital tools can allow for better monitoring and control of the negative impacts of the water systems.

However, while there is an increase of digital adoption in water utilities, the sector still lags behind other industries in integrating new, smart technologies into the built water cycle, creating a largely fragmented market that challenges regulators, utilities and industries.

Industry and utilities around the world should harness digital leapfrog opportunities to increase their efficiency and resilience. In fact, digital tools and the added insights they provide have the potential to enable circular resource use and more reliable and affordable basic water services. Additionally, partnerships across the utility and industry space should be established to improve innovation and integration of new digital technologies and sharing of lessons learned across sectors and geographies would be needed to accelerate the iteration cycle.

Digital tools need to enable transparency, real-time optimisation, mitigation of breakdowns, consistent monitoring, and minimisation of water consumption and losses. Consequently, daily operations need to be assisted by predictive maintenance and automatic process optimisation. By doing so, digital transformation can provide new business models for utilities and industries, for example Data-as-a-Service (DAAS). Industries should build digital offerings around their existing physical components to assist utilities and cities in their digital transformation. Additionally, utilities should regularly interact with and empower customers with data-driven services collected at the point of consumption. Ideally, interactions with regulators should be automated (i.e., based on sensors and algorithms), allowing utilities more time to provide value for the customer. By doing so, digital monitoring tools and utility-industry cooperation will increase accountability for higher performance, and greater levels of engagement will reduce environmental degradation and pollution from point-sources.

The digital transformation will enable seamless integration between utilities and industry, facilitating the optimal use and reuse of resources between stakeholders. To be ready for a 'digital water future', utilities and industry must be enabled by policy and regulation to adopt innovative technologies and ways of working. There needs to be a culture shift internally, which can be supported by third parties, partnerships, or by creating new digital transformation roles, led by champions who are able to drive change. By tapping into the potential of digital water, industry and utilities would ensure more efficient and environmentally friendly operations. Lund et al. (2019) illustrated how a digitally controlled urban drainage system can reduce environmental impact from the sewer system by linking surface water management infrastructure with below ground infrastructure. Li and Wang (2018) showed how a combination of domain knowledge and machine learning can improve water main repair based on predictive methods. Fuertes et al. (2020) showed that a digital twin mirroring a facility can ensure real-time optimisation, avoidance of breakdowns, and planning of staff tasks and maintenance. This transformation is forcing companies to have strict data governance to ensure that data and software is treated at the same level as physical assets.

Additionally, collecting large amounts of data is becoming increasingly important due to the quality of intelligent algorithms. Companies, utilities and regulators will have to partner to get the maximum value from their data, while adhering to the highest privacy and cybersecurity standards. Furthermore, utilities and industry bring different key competencies to the table and will have to harness partnerships in order to solve the digital or domain capacity gap in each individual organisation. To achieve this outcome, there is a need for strategic mobilisation and capacity development of water professionals, through extended partnerships with the Information and Communications Technology (ICT) sector and other leading sectors in the digital transition (i.e., energy and telecommunications sectors).

Decentralised water-wise communities

In emerging markets, most people do not have access to an improved water source or improved WASH services (WHO and UNICEF, 2019). Existing systems are inefficient micro-scale systems (i.e., septic tanks, pit latrines) or ill-managed large systems, both posing a risk to humans and the environment. Centralised systems in emerging markets are often unsustainably financed and ill-equipped to tackle climate change related risks, handle rapid changes in terms of population growth and increased contamination of water resources due to pharmaceutical and agricultural externalities (Camdessus Report, 2003).

Decentralisation faces an uphill battle as regulations and laws need to be reviewed to allow for appropriate standards and compliance monitoring of decentralised approaches (Larsen et al., 2013). On a macro scale, governance issues pose a real problem for decentralised approaches, as centralised systems are preferred due to their prestige and large financial volumes, especially in emerging markets. As smaller, decentralised systems are harder to manage, operate, monitor and oversee, there is a need to eliminate governance hurdles. Furthermore, local governing entities' personnel need to be enabled to assess and decide on infrastructure projects in a sustainable and socially beneficial way.

Integrating digital tools at the local level can support decentralisation (or decentralised approaches) which provides a huge opportunity, especially for emerging markets and growing cities due to lower entry level costs as well as the inherent capacity of flexibility and adaptability to local conditions. Effectively, decentralisation and respective digital solutions enable and increase the ownership by the communities including increased awareness that water is a valuable resource. Additionally, decentralisation and digital solution support job and wealth creation on a local level and save resources by reducing extensive conveyance structures (Larsen et al., 2013).

There is a high potential to achieve sustainable and effective decentralisation by using digital tools to support activities like decentralised rain harvesting, irrigation and drainage for improved local catchment management. Also, well-designed decentralised small treatment systems can be operated and optimised remotely at low costs while using little to no energy for treatment and pumping.

Startups and small and medium-sized enterprises (SMEs) show that piloting new and innovative digital solutions is easier at a decentralised and small scale. Decentralised and intelligently optimised resource recovery enables local reuse, keeping negative as well as positive effects local, and must be an integral part of any activities in local IWRM. Lastly, as the water sector is in dire need of new financing models, the combination of decentralised and digital approaches is likely to harness innovative, small scale, and locally owned financing approaches.

Digital tools for data acquisition and mining, data management, monitoring, and evaluation, as well as new DSS allow for increased cross-sectoral cooperation, involvement of all stakeholders and increased transparency. Moreover, digital tools can support democratisation through increased oversight and therefore increased accountability of decision-makers by tackling prevailing governance issues. Digital tools have the potential to enable management of the complexity of both social and technological dimensions. For example, blockchain technologies could be used to increase the efficiency of an irrigation system as well as increase coordination and trust in communities (Robles et al., 2019).

Accepting the large capacity gap of knowledge in emerging and mature markets, particularly by technical O&M staff, policy and decision-makers, is the first step in finding a solution: an increase of innovative, adaptable on-the-job training and qualification options via digital tools outside of slow-changing traditional structures (universities, institutes) must be driven by various stakeholders (i.e., civil society organisations, individuals, businesses).

Including larger swaths of the society into discussion and decision-making processes will enhance the visibility of the water sector and gather interest by employees and businesses formerly not active in the sector. Decentralised systems will tap into novel small scale financial products, use digital payback schemes as well as community based digital financing solutions to drive access to safe and affordable drinking water and WASH provision.

Lastly, decentralised structures can be more resilient to large scale natural disasters and man- made cybersecurity attacks (Little, 2002; Leigh and Lee, 2019). With the right digital tools, decentralised structures can limit the extent of damage and increase resilience in emergency situations. Best practices will be shared and adapted to local needs by individual decentralised structures. YWPs will need to support the digital transformation of their local communities in this decentralised framework.

Skills for our digital water future

A World Economic Forum Future of Jobs Report stated that '50% of all employees will need reskilling by 2025, as adoption of technology increases' and 'the vast majority of business leaders (94%) now expect employees to pick up new skills on the job' (WEF, 2020). It is therefore clear that there is a need for water professionals to view their training and education in a dynamic light, aiming for ongoing upskilling. Numerous platforms already exist to support this, including but not limited to IWA Learn, Udemy, GetSmarter, FutureLearn, edX and Coursera. Furthermore, universities are increasing their online offerings. This confirms the urgency to adopt and adapt necessary skills and ways of working to achieve our digital water future.

Importantly, the water sector needs to acknowledge the crucial role that professionally trained water professionals and engineers can play in applying the right (digital) solution in the correct step of the water cycle. Technical and engineering skills need to be recognised as fundamental and acquired together with operational management skills to address the technical skills gaps that the industry is currently suffering. Without an in-depth knowledge of the water cycle, there is a risk of investing on the inappropriate technology or applying this in the wrong area.

Nevertheless, technical skills have to be developed together with soft skills (e.g., problem solving, self-management, interpersonal skills) which are also fundamental in adopting new ways of working (agile, scrum, systems thinking, distributed/remote working).

Conclusion
Digital Water is here to stay

Digital Water is undoubtedly the future for the water sector and needs to be considered as a prerequisite in developing and managing the water sector.

The benefits of a digital water sector are clear as we are in dire need of a more sustainable, safe and effective water world. Digitalisation has a role to play in making water affordable, ensuring safe and clean drinking water and protecting water at an ecosystem level, while adapting to a changing climate.

It is evident that there are opportunities for adopting digital tools and ways of working at all levels of water management. Achieving this equitable digital water future requires a commitment to developing partnerships, innovative financing, capacity building and a cohesive strategy. This collective digital water future connects basins, cities, utilities, industry and decentralised water-wise communities with a call for a just and equitable transition, leaving no one behind.

нимание!
асност от

Afterword

The water sector cannot accommodate failure when life depends on reliable access to potable water and safe sanitation. It is understandable that the sector has often taken a more conservative, some may say traditional, approach to innovation and digital developments. However, it is encouraging to see from experiences, examples, lessons and insights shared in this book, that the journey toward a digital water future is underway, albeit at a different pace, rate of uptake and level of maturity across the globe.

In emerging economies, where limited access to resources is often the biggest challenge, the water sector has often taken a backseat to competing and more visible issues, such as housing, transportation, and energy. As a largely underground, out-of-sight system, we only think about water when something goes wrong.

Part of the problem is that we have allowed our behaviour and processes to go unchecked, relying on what worked in the past despite radical changes in the present. Climate change, population growth and urbanisation, as well as human-driven events like pandemics, war and social instability, have accelerated the water crisis. It's no longer "looming"—it's already here.

What we need is a mindset shift regarding how we preserve, use, and reuse water. To quote from Cape Town's Water Strategy underpinned by the principle of a 'Shared Water Future': "It's time to start seeing water as one finite resource. There is no "wastewater", only wasted water."

Smart water management is a critical driver for social and economic security and resilience. This makes it even more imperative to ensure the quality and reliability of the service while simultaneously investing in technologies, human resources, and initiatives to meet future demands.

The stories in these chapters prove that the water industry is ready to rise up to the challenge. There is an eagerness to improve the performance and dynamic resilience of water infrastructure and services using solutions such as Better Information Management, digital twins, and artificial intelligence.

It is understood that every region, utility, and management authority differ in terms of their level of maturity in the journey to a digital water future, but it is clear that we can all learn from each other. It starts with sharing data, knowledge, and experiences to understand what is possible and embracing a progressive, systematic approach to digital transformation.

Ultimately, the message of this book is that everyone stands to benefit from the digital transformation of the water sector. No stakeholder will be left untouched or unchanged and, as such, we are called to embrace the responsibility and right to secure our water resources now and for future generations.

We are at a crossroads where we can either resist the rise of digital solutions and struggle to adapt to water challenges (and global changes), or fully embrace the digital revolution and work with innovators to unlock a new era of water abundance and build a stronger and more economically viable foundation for the future.

It is our hope that this book will trigger a broader conversation about how we can solve water challenges through the application of digital technologies and innovations. We already have a purpose—a felt need for change. Now, we need to be clear on, plan for, and work towards the transformation.

Sheilla De Carvalho
Director Leading Markets, Royal HaskoningDHV

Acknowledgements

The co-editors firstly want to thank the IWA Digital Water Programme's Steering Committee for their conceptualisation, endorsement, and support of this book. It is a compilation of the White Papers produced by dedicated IWA members and water professionals within the water industry and published separately under the IWA Digital Water Programme's White Paper series. We express our gratitude to all of the authors who gave of their time to make this project a success.

We must also thank Clara Martinez and Jacobo Illueca of Idrica, and Jonathan Lavercombe of Stormharvester, for collaborating with us to add the insightful case studies which highlight effective applications of digital tools in water resource management.

We appreciate the comprehensive reviews of Daniel Aguado, Jyoti Gautam, and Marina Batalini de Macedo, and Frank Kizito and Zoran Kapelan for their foreword contributions.

This publication would not have been possible without the impressive design skills of Vivian Langmaack of Ovo Design, who was enthusiastic and committed from the beginning. We thank Vicki Harley for her very detailed and efficient proofreading, ensuring that we present a grammatically sound book.

We recognise and appreciate the affability of IWA Publishing through collaboration with Mark Hammond and Niall Cunniffe, who ensured we produced a strong final document, and could have a successful online and in-person book launch.

Finally, the co-editors would like to thank Erin Jordan, Samuela Guida, Daniel Bemfica, and Kala Vairavamoorthy of the IWA Secretariat, for facilitating this entire process.

Complete bibliography

Chapter 1

Affleck, M., Allen, G., Bozalongo, C.B., Brown, H., Gasson, C., Gasson, J., González-Manchón, C., Hudecova, M., McFie, A., Shuttleworth, H., Tan, M., Underwood, B., Uzelac, J., Virgili, F., and Walker, C., 2016. Water's digital future. Global Water Intelligence.

Bailey, J., Harris, E., Keedwell, E., Djordjevic, S., Kapelan, Z., 2016. Developing Decision Tree Models to Create a Predictive Blockage Likelihood Model for Real-World Wastewater Networks. Procedia Eng. 154, 1209–1216. https://doi.org/10.1016/j.proeng.2016.07.433

Biswas, B., 2016. Is Blockchain a Reality of an Innovation Wish List in Utilities [WWW Document]. URL https://www.capgemini.com/2017/11/is-blockchain- a-reality-or-an-innovation-wish-list-in-utilities/# (accessed 4.29.19).

CISION PR Newswire, 2018. Global Smart Water Leakage Management Solutions Market Report 2018-2025 [WWW Document]. Mark. Rep. URL https://www.prnewswire.com/news-releases/global-smart-water-leakage-management-solutions-market-report-2018-2025-300737280.html (accessed 4.29.19).

Cohen, L., 2018. Augmented and Virtual Reality in Operations A guide for investment. Capgemini Res. Inst.

Eggers, W.D., Skowron, J., 2018. Forces of Change: Smart Cities [WWW Document]. Web Artic. URL https://www2.deloitte.com/insights/ us/en/focus/smart-city/overview.html (accessed 4.29.19).

Falco, G.J., Webb, W.R., 2015. Water Microgrids: The Future of Water Infrastructure Resilience. Procedia Eng. 118, 50–57. https://doi.org/10.1016/j.proeng.2015.08.403

Gartner, 2018. Gartner Survey Shows Organizations Are Slow to Advance in Data and Analytics [WWW Document]. Web Artic. URL https://www.gartner.com/ en/newsroom/press-releases/2018-02-05-gartner-survey-shows- organizations-are-slow-to-advance-in-data-and-analytics (accessed 4.29.22).

Grievson, O., 2017. Smart Wastewater Networks, From Micro To Macro [WWW Document]. Web Artic. URL https://www.wateronline.com/doc/smart-wastewater-networks-from-micro-to-macro-0001 (accessed 4.29.19).

Hagel, J., Seely-Brown, J., Samoylova, T., Lui, M., 2013. From exponential technologies to exponential innovation. Netherlands, Denmark.

Hexagon, 2018. Innovative Ways of Working: Digital Twins in the U.K. Water Industry [WWW Document]. Blog post. URL https://blog.hexagonsafetyinfrastructure.com/innovative-ways-of-working-digital-twins-in-the-u-k-water-industry/ (accessed 4.29.19).

Hill, T., 2018. How Artificial Intelligence is Reshaping the Water Sector [WWW Document]. Web Artic. URL https://waterfm.com/artificial-intelligence-reshaping-water-sector/#:~:text=Because of AI's ability to,plan for the years ahead. (accessed 4.29.19).

IWA, 2019. Digital Water Programme [WWW Document]. Web Artic.

Kane, G., Phillips, A., Copulsky, J., Andrus, G., 2019. The duct tape guide to digital strategy. Netherlands, Denmark.

Kelly, E., 2015. Introduction: Business ecosystems come of age. Netherlands, Denmark.

Ministry of Environment Denmark, 2019. Identify sources of water loss [WWW Document]. Web Page. URL https://eng.mst.dk/nature-water/water-at-home/water-loss/ (accessed 4.29.19).

South China Morning Post, 2019. Offshore Water Quality in Shenzhen Worst of Guangdong's Costal Cities [WWW Document]. Web Artic.

The Nation, 2018. Digitalisation Can Solve Water and Climate Crisis [WWW Document]. Web Artic. URL http://www.nationmultimedia.com/ detail/opinion/30343725 (accessed 4.29.19).

United Nations, 2018. 68% of the World Population Projected to Live in Urban Areas by 2050, Says UN [WWW Document]. Web Artic. URL https://www.un.org/development/desa/en/news/population/2018-revision-of-world-urbanization-prospects.html (accessed 4.29.19).

University of Denver, 2015. Non-Revenue Water Loss: the Invisible Global Problem [WWW Document]. Univ. Denver Water Law Rev. 2.

Veolia, 2017. Digital to the Rescue: Making Water Management Smarter in City Networks [WWW Document]. Web Artic. URL https://www.veolia.com/anz/rethinking-sustainability/rethinking-sustainability-blog/digital-rescue- making-water-management (accessed 4.29.19).

vGIS, 2019. Augmented reality for utilities surges ahead: Toronto Water pilots holographic GIS [WWW Document]. Web Artic. URL https://www.vgis.io/2018/01/24/augmented-reality-utilities-surges-ahead-toronto-water-pilots-holographic-gis/ (accessed 4.29.22).

Chapter 2

Blokker, M., Agudelo-Vera, C., Moerman, A., van Thienen, P., Pieterse-Quirijns, I., 2017. Review of applications for SIMDEUM, a stochastic drinking water demand model with a small temporal and spatial scale. Drink. Water Eng. Sci. 10, 1–12. https://doi.org/10.5194/dwes-10-1-2017

Campisano, A., Ple, J.C., Muschalla, D., Pleau, M., Vanrolleghem, P.A., 2013. Potential and limitations of modern equipment for real time control of urban wastewater systems. Urban Water J. 10, 300–311. https://doi.org/10.1080/1573062X.2013.763996

Conjeos, P., 2020. A case study of the Digital Twin Approach taken with the city of Valencia.

CRC for water sensitive cities, 2019. Water Sensitive Cities Scenario Tool: utilise water sensitive city outcomes in scenario planning [WWW Document]. Scenar. tool. URL https://watersensitivecities.org.au/solutions/water-sensitive-cities-scenario-tool/ (accessed 9.24.20).

de Souza Groppo, G., Costa, M.A., Libânio, M., 2019. Predicting water demand: a review of the methods employed and future possibilities. Water Supply 19, 2179–2198. https://doi.org/10.2166/ws.2019.122

Ebbers, M., Abdel-Gayed, A., Budhi, V.B., Dolot, F., Kamat, V., Picone, R., Trevelin, J., 2013. Addressing Data Volume, Velocity, and Variety with IBM InfoSphere Streams V3.0, 1st ed. IBM redbooks.

Ellison, M., 2016. Smart flood alleviation system protects Portsmouth. Mag. Artic. 2.

H2O-Online, 2019. Toekomstverkenning:alternatievebronnenvoor drinkwater in Nederland [WWW Document]. Web Artic. URL https://www.h2owaternetwerk.nl/images/2019/augustus/H2O-Online_190828_Alternatieve_drinkwaterbronnen.pdf (accessed 9.29.19).

Hauduc, H., Gillot, S., Rieger, L., Ohtsuki, T., Shaw, A., Takács, I., Winkler, S., 2009. Activated sludge modelling in practice: an international survey. Water Sci. Technol. 60, 1943–1951. https://doi.org/10.2166/wst.2009.223

IPCC, 2007. Guidance Notes for Lead Authors of the IPCC Fourth Assessment Report on Addressing Uncertainties.

ISO, 2003. ISO 15839:2003: Water quality — On-line sensors/analysing equipment for water — Specifications and performance tests.

IWA, 2019. Digital Water Programme [WWW Document]. Web Artic.

Jiang, J., Tang, S., Han, D., Fu, G., Solomatine, D., Zheng, Y., 2020. A comprehensive review on the design and optimization of surface water quality monitoring networks. Environ. Model. Softw. 132, 104792. https://doi.org/10.1016/j.envsoft.2020.104792

Kurgan, L.A., Musilek, P., 2006. A survey of Knowledge Discovery and Data Mining process models. Knowl. Eng. Rev. 21, 1–24. https://doi.org/10.1017/S0269888906000737

Kwakkel, J., 2017. Managing deep uncertainty: Exploratory modeling, adaptive plans and joint sense making [WWW Document]. Web Artic. URL https://www.deepuncertainty.org/2017/10/15/managing-deep-uncertainty-exploratory-modeling-adaptive-plans-and-joint-sense-making/ (accessed 12.9.19).

Kwakkel, J.H., Haasnoot, M., Walker, W.E., 2016. Comparing Robust Decision-Making and Dynamic Adaptive Policy Pathways for model-based decision support under deep uncertainty. Environ. Model. Softw. 86, 168–183. https://doi.org/10.1016/j.envsoft.2016.09.017

Maiolo, M., Mendicino, G., Pantusa, D., Senatore, A., 2017. Optimization of Drinking Water Distribution Systems in Relation to the Effects of Climate Change. Water 9. https://doi.org/10.3390/w9100803

Masson-Delmotte, V.P., Zhai, H.-O., Pörtner, D., Roberts, J., Skea, P.R., Shukla, A., Pirani, W., Moufouma-Okia, C., Péan, R., Pidcock, S., Connors, J.B.R., Matthews, Y., Chen, X., Zhou, M.I., Gomis, E., Lonnoy, T., Maycock, M., Tignor, Waterfield, T., 2018. Summary for Policymakers. In: Global Warming of 1.5°C. An IPCC Special Report on the impacts of global warming of 1.5°C above pre-industrial levels and related global greenhouse gas emission pathways, in the context of strengthening the global response to. New York, USA. https://doi.org/10.1017/9781009157940.001

Morley, M., Savić, D., 2020. Water Resource Systems Analysis for Water Scarcity Management: The Thames Water Case Study. Water 12. https://doi.org/10.3390/w12061761

New Climate Analytics, 2019. Climate action tracker [WWW Document].

Nikolopoulos, D., van Alphen, H.-J., Vries, D., Palmen, L., Koop, S., van Thienen, P., Medema, G., Makropoulos, C., 2019. Tackling the "New Normal": A Resilience Assessment Method Applied to Real-World Urban Water Systems. Water 11. https://doi.org/10.3390/w11020330

Normandeau, K., 2013. Beyond Volume, Variety and Velocity is the issue of big Data Veracity [WWW Document]. Web Artic. URL http://insidebigdata.com/2013/09/12/beyond-volume-variety-velocity-issue-big-data-veracity

Ohmura, K., Thürlimann, C.M., Kipf, M., Carbajal, J.P., Villez, K., 2019. Characterizing long-term wear and tear of ion-selective pH sensors. Water Sci. Technol. 80, 541–550. https://doi.org/10.2166/wst.2019.301

Pidcock, R., Pearce, R., McSweeney, R., 2019. Mapped: How climate change affects extreme weather around the world [WWW Document]. Interact. map. URL https://www.carbonbrief.org/mapped-how-climate-change-affects-extreme-weather-around-the-world/ (accessed 12.9.19).

Rieger, L., Takács, I., Villez, K., Siegrist, H., Lessard, P., Vanrolleghem, P.A., Comeau, Y., 2010. Data reconciliation for wastewater treatment plant simulation studies-planning for high-quality data and typical sources of errors. Water Environ. Res. a Res. Publ. Water Environ. Fed. 82, 426–433. https://doi.org/10.2175/106143009x12529484815511

Sagiroglu, S., Sinanc, D., 2013. Big data: A review, in: 2013 International Conference on Collaboration Technologies and Systems (CTS). pp. 42–47. https://doi.org/10.1109/CTS.2013.6567202

Sardinha, J., Serranito, F., Donnelly, A., Marmelo, V., Saraiva, P., Dias, N., Guimarães, R., Morais, D., Rocha, V., 2017. Active water loss control. Portugal.

Savic, D.A., Morley, M.S., Khoury, M., 2016. Serious Gaming for Water Systems Planning and Management. Water 8. https://doi.org/10.3390/w8100456

SIM4NEXUS, 2020. SIM4NEXUS: water-energy-land-food and climate nexus for resource efficiency [WWW Document]. Web Page. URL https://sim4nexus.eu/ (accessed 3.5.20).

Stofberg, S., Hofman-Caris, R., Pronk, G., Alphen, en H.-J. van, Putters, B., 2019. Toekomstverkenning: alternatieve bronnen voor drinkwater in Nederland. Flanders, Netherlands.

UKWIR, 2013. The use of active system control when designing sewerage schemes - a guide. London, UK.

Chapter 3

Aguado, D., Blumensaat, F., Antonio Baeza, J., Villez, K., Ruano Victoria, M., Samuelsson, O., Plana, Q., 2021. The value of meta-data for water resource recovery facilities. IWA Digit. Water White Pap. Ser. 15.

Ahmed, W., Angel, N., Edson, J., Bibby, K., Bivins, A., O'Brien, J.W., Choi, P.M., Kitajima, M., Simpson, S.L., Li, J., Tscharke, B., Verhagen, R., Smith, W.J.M., Zaugg, J., Dierens, L., Hugenholtz, P., Thomas, K. V, Mueller, J.F., 2020. First confirmed detection of SARS-CoV-2 in untreated wastewater in Australia: A proof of concept for the wastewater surveillance of COVID-19 in the community. Sci. Total Environ. 728, 138764. https://doi.org/10.1016/j.scitotenv.2020.138764

Allamano, P., Croci, A., Laio, F., 2015. Toward the camera rain gauge. Water Resour. Res. 51, 1744–1757. https://doi.org/10.1002/2014WR016298

Babovic, V., Cañizares, R., Jensen, H.R., Klinting, A., 2001. Neural networks as routine for error updating of numerical models. J. Hydraul. Eng. 127, 181–193.

Babovic, V., Drécourt, J.-P., Keijzer, M., Friss Hansen, P., 2002. A data mining approach to modelling of water supply assets. Urban Water 4, 401–414. https://doi.org/10.1016/S1462-0758(02)00034-1

Borzooei, S., Miranda, G.H.B., Abolfathi, S., Scibilia, G., Meucci, L., Zanetti, M.C., 2020. Application of unsupervised learning and process simulation for energy optimization of a WWTP under various weather conditions. Water Sci. Technol. 81, 1541–1551. https://doi.org/10.2166/wst.2020.220

Butler, D., Farmani, R., Fu, G., Ward, S., Diao, K., Astaraie-Imani, M., 2014. A new approach to urban water management: safe and sure. Procedia Eng. 89, 347–354. https://doi.org/10.1016/j.proeng.2014.11.198

Conlin, H., O'meara, J., 2006. Assessing Longford Gas Plant 1 Staffing Arrangements. Process Saf. Environ. Prot. 84, 10–20. https://doi.org/10.1205/psep.04245

Corominas, L., Garrido-Baserba, M., Villez, K., Olsson, G., Cortés, U., Poch, M., 2018. Transforming data into knowledge for improved wastewater treatment operation: a critical review of techniques. Environ. Model. Softw. 106, 89–103. https://doi.org/10.1016/j.envsoft.2017.11.023

Crits-Christoph, A., Kantor, R.S., Olm, M.R., Whitney, O.N., Al-Shayeb, B., Lou, Y.C., Flamholz, A., Kennedy, L.C., Greenwald, H., Hinkle, A., Hetzel, J., Spitzer, S., Koble, J., Tan, A., Hyde, F., Schroth, G., Kuersten, S., Banfield, J.F., Nelson, K.L., Pettigrew, M.M., 2021. Genome Sequencing of Sewage Detects Regionally Prevalent SARS-CoV-2 Variants. MBio 12, e02703-20. https://doi.org/10.1128/mBio.02703-20

Daigger, G.T., 2009. Evolving urban water and residuals management paradigms: water reclamation and reuse, decentralization, and resource recovery. Water Environ. Res. a Res. Publ. Water Environ. Fed. 81, 809–823. https://doi.org/10.2175/106143009x425898

Economou, T., Bailey, T.C., Kapelan, Z., 2014. MCMC implementation for Bayesian hidden semi-Markov models with illustrative applications. Stat. Comput. 24, 739–752. https://doi.org/10.1007/s11222-013-9399-z

Fischer, E.M., Sedláček, J., Hawkins, E., Knutti, R., 2014. Models agree on forced response pattern of precipitation and temperature extremes. Geophys. Res. Lett. 41, 8554–8562. https://doi.org/10.1002/2014GL062018

Garrido-Baserba, M., Corominas, L., Cortés, U., Rosso, D., Poch, M., 2020. The Fourth-Revolution in the Water Sector Encounters the Digital Revolution. Environ. Sci. Technol. 54, 4698–4705. https://doi.org/10.1021/acs.est.9b04251

Grievson, O., 2020. Digital water: the role of instrumentation in digital transformation. IWA Digit. water White Pap. Ser. 19.

Guest, J.S., Skerlos, S.J., Barnard, J.L., Beck, M.B., Daigger, G.T., Hilger, H., Jackson, S.J., Karvazy, K., Kelly, L., Macpherson, L., Mihelcic, J.R., Pramanik, A., Raskin, L., Van Loosdrecht, M.C.M., Yeh, D., Love, N.G., 2009. A New Planning and Design Paradigm to Achieve Sustainable Resource Recovery from Wastewater. Environ. Sci. \& Technol. 43, 6126–6130. https://doi.org/10.1021/es9010515

Gwenzi, W., 2022. Wastewater, waste, and water-based epidemiology (WWW-BE): A novel hypothesis and decision-support tool to unravel COVID-19 in low-income settings? Sci. Total Environ. 806, 150680. https://doi.org/10.1016/j.scitotenv.2021.150680

Harper, D., 2019. Online Etymology Dictionary: Resilience [WWW Document]. Web Page. URL https://www.etymonline.com/word/resilience (accessed 11.17.19).

Holling, C.S., 1973. Resilience and Stability of Ecological Systems 4, 4050-. https://doi.org/10.1146/annurev.es.04.110173.000245

Holloway, T. G, Jeppsson, U., Fu, G., Secu, S., Molinos-Senante, M., 2021. Modelling wastewater treatment resilience for improved decision making and resource recovery. IWA.

Holloway, T G., Williams, J.B., Ouelhadj, D., Yang, G., 2021. Dynamic resilience for biological wastewater treatment processes: Interpreting data for process management and the potential for knowledge discovery. J. Water Process Eng. 42, 102170. https://doi.org/10.1016/j.jwpe.2021.102170

IWA, 2019. Digital Water Programme [WWW Document]. Web Artic.

Jiang, S., Babovic, V., Zheng, Y., Xiong, J., 2019. Advancing Opportunistic Sensing in Hydrology: A Novel Approach to Measuring Rainfall With Ordinary Surveillance Cameras. Water Resour. Res. 55, 3004–3027. https://doi.org/10.1029/2018WR024480

Juan-García, P., Butler, D., Comas, J., Darch, G., Sweetapple, C., Thornton, A., Corominas, L., 2017. Resilience theory incorporated into urban wastewater systems management. State of the art. Water Res. 115, 149–161. https://doi.org/10.1016/j.watres.2017.02.047

Karmous-Edwards, G., Conejos, P., Mahinthakumar, K., Braman, S., Vicat-Blanc, P., Barba, J., 2019. Foundations for building a digital twin for water utilities.

Kehrein, P., van Loosdrecht, M., Osseweijer, P., Garfí, M., Dewulf, J., Posada, J., 2020. A critical review of resource recovery from municipal wastewater treatment plants — market supply potentials{,} technologies and bottlenecks. Environ. Sci. Water Res. Technol. 6, 877–910. https://doi.org/10.1039/C9EW00905A

Larsen, D.A., Green, H., Collins, M.B., Kmush, B.L., 2021. Wastewater monitoring, surveillance and epidemiology: a review of terminology for a common understanding. FEMS Microbes 2. https://doi.org/10.1093/femsmc/xtab011

Mao, K., Zhang, K., Du, W., Ali, W., Feng, X., Zhang, H., 2020. The potential of wastewater-based epidemiology as surveillance and early warning of infectious disease outbreaks. Curr. Opin. Environ. Sci. Heal. 17, 1–7. https://doi.org/10.1016/j.coesh.2020.04.006

Mugume, S.N., Gomez, D.E., Fu, Guangtao, Farmani, R., Butler, D., 2015. A global analysis approach for investigating structural resilience in urban drainage systems. Water Res. 81, 15–26. https://doi.org/10.1016/j.watres.2015.05.030

Myrans, J., Everson, R., Kapelan, Z., 2018. Automated detection of faults in sewers using CCTV image sequences. Autom. Constr. 95, 64–71. https://doi.org/10.1016/j.autcon.2018.08.005

NASA, N.A. and S.A., 2016. Scientific consensus: Earth's climate is warming [WWW Document]. Web Page. URL https://climate.nasa.gov/scientific-consensus

Newhart, K.B., Holloway, R.W., Hering, A.S., Cath, T.Y., 2019. Data-driven performance analyses of wastewater treatment plants: a review. Water Res. 157, 498–513. https://doi.org/10.1016/j.watres.2019.03.030

Ofwat, 2019. Ofwat: PR19 Initial assessment of business plans – key documents [WWW Document]. Web Resour. page. URL https://www.ofwat.gov.uk/investor/pr19/ (accessed 6.18.19).

Pedersen, A.N., Borup, M., Brink-Kjær, A., Christiansen, L.E., Mikkelsen, P.S., 2021. Living and Prototyping Digital Twins for Urban Water Systems: Towards Multi-Purpose Value Creation Using Models and Sensors. Water 13. https://doi.org/10.3390/w13050592

Poch, M., Garrido-Baserba, M., Corominas, L., Perelló-Moragues, A., Monclús, H., Cermerón-Romero, M., Melitas, N., Jiang, S.C., Rosso, D., 2020. When the fourth water and digital revolution encountered COVID-19. Sci. Total Environ. 744, 140980. https://doi.org/10.1016/j.scitotenv.2020.140980

Pörtner, H.O., Roberts, D.C., Tignor, M., Poloczanska, E.S., Mintenbeck, A., Alegría, A., Craig, M., Langsdorf, S., Löschke, S., Möller, V., Okem, A., Rama, B., 2022. IPCC, 2022: Climate Change 2022: Impacts, Adaptation, and Vulnerability. Contribution of Working Group II to the Sixth Assessment Report of the Intergovernmental Panel on Climate Change. Cambridge, UK.

Ramos, G., Hynes, W., 2020. A systematic resilience approach to dealing with COVID-19 and future shocks.

Romano, M., Kapelan, Z., 2014. Adaptive water demand forecasting for near real-time management of smart water distribution systems. Environ. Model. Softw. 60, 265–276. https://doi.org/10.1016/j.envsoft.2014.06.016

Romano, M., Kapelan, Z., Savić, D.A., 2014. Automated Detection of Pipe Bursts and Other Events in Water Distribution Systems. J. Water Resour. Plan. Manag. 140, 457–467. https://doi.org/10.1061/(asce)wr.1943-5452.0000339

Sarni, W., White, C., Webb, R., Cross, K., Glotzbach, R., 2019. Digital Water Industry leaders chart the transformation journey (No. 1), 1. London, UK.

Stentoft, P.A., 2020. Stochastic Modelling and Predictive Control of Wastewater Treatment Processes. Technical University of Denmark.

Sweetapple, C., Fu, G., Farmani, R., Butler, D., 2022a. General resilience: Conceptual formulation and quantitative assessment for intervention development in the urban wastewater system. Water Res. 211, 118108. https://doi.org/10.1016/j.watres.2022.118108

Sweetapple, C., Melville-Shreeve, P., Chen, A.S., Grimsley, J.M.S., Bunce, J.T., Gaze, W., Fielding, S., Wade, M.J., 2022b. Building knowledge of university campus population dynamics to enhance near-to-source sewage surveillance for SARS-CoV-2 detection. Sci. Total Environ. 806, 150406. https://doi.org/10.1016/j.scitotenv.2021.150406

Tulchinsky, T.H., 2018. John Snow, Cholera, the Broad Street Pump; Waterborne Diseases Then and Now. Case Stud. Public Heal. https://doi.org/10.1016/B978-0-12-804571-8.00017-2

Wade, M.J., Lo Jacomo, A., Armenise, E., Brown, M.R., Bunce, J.T., Cameron, G.J., Fang, Z., Farkas, K., Gilpin, D.F., Graham, D.W., Grimsley, J.M.S., Hart, A., Hoffmann, T., Jackson, K.J., Jones, D.L., Lilley, C.J., McGrath, J.W., McKinley, J.M., McSparron, C., Nejad, B.F., Morvan, M., Quintela-Baluja, M., Roberts, A.M.I., Singer, A.C., Souque, C., Speight, V.L., Sweetapple, C., Walker, D., Watts, G., Weightman, A., Kasprzyk-Hordern, B., 2022. Understanding and managing uncertainty and variability for wastewater monitoring beyond the pandemic: Lessons learned from the United Kingdom national COVID-19 surveillance programmes. J. Hazard. Mater. 424, 127456. https://doi.org/10.1016/j.jhazmat.2021.127456

Walker, B., Holling, C.S., Carpenter, S.R., Kinzig, A., 2004. Resilience, Adaptability and Transformability in Social– ecological Systems. Ecol. Soc. 9, 5.

Wise on Water, 2020. Urban resilience: UK municipalities, water utilities collaborate to combat effects of climate change. Web Artic. 3.

Wu, J., Wang, Z., Lin, Y., Zhang, L., Chen, J., Li, P., Liu, W., Wang, Y., Yao, C., Yang, K., 2021. Technical framework for wastewater-based epidemiology of SARS-CoV-2. Sci. Total Environ. 791, 148271. https://doi.org/10.1016/j.scitotenv.2021.148271

Chapter 4

Acker, J., Soebiyanto, R., Kiang, R., Kempler, S., 2014. Use of the NASA Giovanni Data System for Geospatial Public Health Research: Example of Weather-Influenza Connection. ISPRS Int. J. Geo-Information 3, 1372–1386. https://doi.org/10.3390/ijgi3041372

Bar-Or, I., Yaniv, K., Shagan, M., Ozer, E., Erster, O., Mendelson, E., Mannasse, B., Shirazi, R., Kramarsky-Winter, E., Nir, O., Abu-Ali, H., Ronen, Z., Rinott, E., Lewis, Y.E., Friedler, E., Bitkover, E., Paitan, Y., Berchenko, Y., Kushmaro, A., 2020. Regressing SARS-CoV-2 sewage measurements onto COVID-19 burden in the population: a proof-of-concept for quantitative environmental surveillance. medRxiv. https://doi.org/10.1101/2020.04.26.20073569

Bauer, G.K., 2019. Mobile Technology Is Driving the Digitisation of Water Utilities in Emerging Markets. London, UK.

Business & Sustainable Development Commission, 2017. Better business better world. London, UK.

CGAP, 2019. Testing the Waters: Digital Payments for Water and Sanitation [WWW Document]. Web Artic. URL https://www.cgap.org/research/publication/testing-waters-digital-payments-water-and-sanitation

Falman, J.C., Fagnant-Sperati, C.S., Kossik, A.L., Boyle, D.S., Meschke, J.S., 2019. Evaluation of Secondary Concentration Methods for Poliovirus Detection in Wastewater. Food Environ. Virol. 11, 20–31. https://doi.org/10.1007/s12560-018-09364-y

Gandhi, M., Yokoe, D.S., Havlir, D. V, 2020. Asymptomatic Transmission, the Achilles' Heel of Current Strategies to Control Covid-19. N. Engl. J. Med. 382, 2158–2160. https://doi.org/10.1056/NEJMe2009758

Gatidou, G., Kinyua, J., van Nuijs, A.L.N., Gracia-Lor, E., Castiglioni, S., Covaci, A., Stasinakis, A.S., 2016. Drugs of abuse and alcohol consumption among different groups of population on the Greek Island of Lesvos through sewage-based epidemiology. Sci. Total Environ. 563–564, 633–640. https://doi.org/10.1016/j.scitotenv.2016.04.130

Gracia-Lor, E., Castiglioni, S., Bade, R., Been, F., Castrignanò, E., Covaci, A., González-Mariño, I., Hapeshi, E., Kasprzyk-Hordern, B., Kinyua, J., Lai, F.Y., Letzel, T., Lopardo, L., Meyer, M.R., O'Brien, J., Ramin, P., Rousis, N.I., Rydevik, A., Ryu, Y., Santos, M.M., Senta, I., Thomaidis, N.S., Veloutsou, S., Yang, Z., Zuccato, E., Bijlsma, L., 2017. Measuring biomarkers in wastewater as a new source of epidemiological information: Current state and future perspectives. Environ. Int. 99, 131–150. https://doi.org/10.1016/j.envint.2016.12.016

GSMA, 2020. Digital solutions for the urban poor.

GSMA, 2015. GSMA | Wonderkid: Digitising Water Utilities in Kenya | Mobile for Development, Report. Kenya, Africa.

Kocamemi, B.A., Kurt, H., Hac\ioglu, S., Yaral\i, C., Saatci, A.M., Pakdemirli, B., 2020. First Data-Set on SARS-CoV-2 Detection for Istanbul Wastewaters in Turkey. medRxiv. https://doi.org/10.1101/2020.05.03.20089417

Lorenzo, M., Picó, Y., 2019. Wastewater-based epidemiology: current status and future prospects. Curr. Opin. Environ. Sci. Heal. 9, 77–84. https://doi.org/10.1016/j.coesh.2019.05.007

Lu, F.S., Nguyen, A.T., Link, N.B., Lipsitch, M., Santillana, M., 2020. Estimating the Early Outbreak Cumulative Incidence of COVID-19 in the United States: Three Complementary Approaches. medRxiv. https://doi.org/10.1101/2020.04.18.20070821

PATH, 2020. New tool detects poliovirus hiding in the environment [WWW Document]. Web Artic. URL https://www.path.org/articles/new-tool-detects-poliovirus-hiding-environment/

Randazzo, W., Truchado, P., Cuevas-Ferrando, E., Simón, P., Allende, A., Sánchez, G., 2020. SARS-CoV-2 RNA in wastewater anticipated COVID-19 occurrence in a low prevalence area. Water Res. 181, 115942. https://doi.org/10.1016/j.watres.2020.115942

Sarni, W., White, C., Webb, R., Cross, K., Glotzbach, R., 2019. Digital Water Industry leaders chart the transformation journey (No. 1), 1. London, UK.

van Zyl, W.B., Zhou, N.A., Wolfaardt, M., Matsapola, P.N., Ngwana, F.B., Symonds, E.M., Fagnant-Sperati, C.S., Shirai, J.H., Kossik, A.L., Beck, N.K., Komen, E., Mwangi, B., Nyangao, J., Boyle, D.S., Borus, P., Taylor, M.B., Scott Meschke, J., 2019. Detection of potentially pathogenic enteric viruses in environmental samples from Kenya using the bag-mediated filtration system. Water Supply 19, 1668–1676. https://doi.org/10.2166/ws.2019.046

WHO, 2020a. Strategic Preparedness and Response Plan for the Novel Coronavirus.

WHO, 2020b. COVID-19 Strategy Update.

Wu, F., Zhang, J., Xiao, A., Gu, X., Lee, W.L., Armas, F., Kauffman, K., Hanage, W., Matus, M., Ghaeli, N., Endo, N., Duvallet, C., Poyet, M., Moniz, K., Washburne, A.D., Erickson, T.B., Chai, P.R., Thompson, J., Alm, E.J., Gilbert, J.A., 2020. SARS-CoV-2 Titers in Wastewater Are Higher than Expected from Clinically Confirmed Cases. mSystems 5, e00614-20. https://doi.org/10.1128/mSystems.00614-20

Wurtzer, S., Marechal, V., Mouchel, J.M., Maday, Y., Teyssou, R., Richard, E., Almayrac, J.L., Moulin, L., 2020. Evaluation of lockdown impact on SARS-CoV-2 dynamics through viral genome quantification in Paris wastewaters. medRxiv. https://doi.org/10.1101/2020.04.12.20062679

Zhou, N.A., Fagnant-Sperati, C.S., Shirai, J.H., Sharif, S., Zaidi, S.Z., Rehman, L., Hussain, J., Agha, R., Shaukat, S., Alam, M., Khurshid, A., Mujtaba, G., Salman, M., Safdar, R.M., Mahamud, A., Ahmed, J., Khan, S., Kossik, A.L., Beck, N.K., Matrajt, G., Asghar, H., Bandyopadhyay, A.S., Boyle, D.S., Meschke, J.S., 2018. Evaluation of the bag-mediated filtration system as a novel tool for poliovirus environmental surveillance: Results from a comparative field study in Pakistan. PLoS One 13, 1–15. https://doi.org/10.1371/journal.pone.0200551

Chapter 5

Bordel, B., Martin, D., Alcarria, R., Robles, T., 2019. A Blockchain-based Water Control System for the Automatic Management of Irrigation Communities, in: 2019 IEEE International Conference on Consumer Electronics (ICCE). pp. 1–2. https://doi.org/10.1109/ICCE.2019.8661940

Burdett, R., Rode, P., 2018. Shaping cities in an urban age, 1st ed. Phaidon, London, UK.

Camdassus, M., 2003. Report of the World Panel on Financing Water Infrastructure. Johannesburg, Southa Africa.

Conejos Fuertes, P., Martínez Alzamora, F., Hervás Carot, M., Alonso Campos, J.C., 2020. Building and exploiting a Digital Twin for the management of drinking water distribution networks. Urban Water J. 17, 704–713. https://doi.org/10.1080/1573062X.2020.1771382

de Souza Groppo, G., Costa, M.A., Libânio, M., 2019. Predicting water demand: a review of the methods employed and future possibilities. Water Supply 19, 2179–2198. https://doi.org/10.2166/ws.2019.122

European Commision, 2020. Smart Cities and Communities [WWW Document]. Web Artic. URL https://digital-strategy.ec.europa.eu/en/policies/smart-cities-and-communities (accessed 8.15.22).

Guida, S., 2018. Industry leaders chart the transformation journey [WWW Document]. Web Artic. URL https://iwa-network.org/projects/digital-water-report/ (accessed 8.15.22).

Ho, J.C., Michalak, A.M., Pahlevan, N., 2019. Widespread global increase in intense lake phytoplankton blooms since the 1980s. Nature 574, 667–670. https://doi.org/10.1038/s41586-019-1648-7

IWA, 2018. Wastewater Report: the reuse opportunity. London UK.

Jones, E., van Vliet, M., Qadir, M., Bierkens, M., 2021. Country-level and gridded estimates of wastewater production, collection, treatment and reuse. Earth Syst. Sci. Data 237–254. https://doi.org/10.5194/essd-13-237-2021, 2021

Larsen, T.A., Udert, K.M., Lienert, J., 2013. Source Separation and Decentralization for Wastewater Management. IWA Publishing. https://doi.org/10.2166/9781780401072

Leigh, N.G., Lee, H., 2019. Sustainable and Resilient Urban Water Systems: The Role of Decentralization and Planning. Sustain. . https://doi.org/10.3390/su11030918

Li, Z., Wang, Y., 2018. Domain Knowledge in Predictive Maintenance for Water Pipe Failures, in: Zhou, J., Chen, F. (Eds.), Human and Machine Learning: Visible, Explainable, Trustworthy and Transparent. Springer International Publishing, Cham, pp. 437–457. https://doi.org/10.1007/978-3-319-90403-0_21

Little, R.G., 2002. Controlling Cascading Failure: Understanding the Vulnerabilities of Interconnected Infrastructures. J. Urban Technol. 9, 109–123. https://doi.org/10.1080/106307302317379855

Lund, N.S.V., Borup, M., Madsen, H., Mark, O., Arnbjerg-Nielsen, K., Mikkelsen, P.S., 2019. Integrated stormwater inflow control for sewers and green structures in urban landscapes. Nat. Sustain. 2, 1003–1010. https://doi.org/10.1038/s41893-019-0392-1

Sarni, W., White, C., Webb, R., Cross, K., Glotzbach, R., 2019. Digital Water Industry leaders chart the transformation journey (No. 1), 1. London, UK.

Serranito, F., Donnelly, A., 2015. Controlo Ativo de Perdas de Água, 1st ed. Empresa Portuguesa das Águas Livres S.A, Lisbon, Portugal.

UN, 2017. Water and ecosystems: challenges and opportunities [WWW Document]. Web Artic. URL https://www.unwater.org/water-facts/ecosystems/ (accessed 8.15.22).

UN Environment Program, 2017. Annual report.

UN World Water Assessment Programme, 2009. The United Nations World Water Development Report 3. London, UK.

Water World Magazine, 2017. Cyber Security: How Water Utilities Can Protect Against Threats [WWW Document]. Magazine. URL https://www.waterworld.com/home/article/16201183/cyber-security-how-water-utilities-can-protect-against-threats (accessed 8.15.22).

WEF, 2020a. These are the top 10 job skills of tomorrow – and how long it takes to learn them [WWW Document]. Web Artic. URL https://www.weforum.org/agenda/2020/10/top-10-work-skills-of-tomorrow-how-long-it-takes-to-learn-them/ (accessed 8.15.22).

WEF, 2020b. The Global Risks Report 2020: 15th edition. Geneva, Switzerland.

WHO, 2019. 1 in 3 people globally do not have access to safe drinking water – UNICEF, WHO [WWW Document]. Web Artic. URL https://www.who.int/news-room/detail/18-06-2019-1-in-3-people-globally-do-not-have-access-to-safe-drinking-water-unicef-who

Wright, H., Williams, S., Hargrave, J., zu Dohna, F., 2018. Cities Alive: designing for urban childhoods. London, UK.

Yñiguez, A.T., Ottong, Z.J., 2020. Predicting fish kills and toxic blooms in an intensive mariculture site in the Philippines using a machine learning model. Sci. Total Environ. 707, 136173. https://doi.org/10.1016/j.scitotenv.2019.136173

IWA Publishing's authorised EU representative for General Product Safety Regulations is Diane D'Arras, 15 rue Duret, 75116 Paris, France, e-mail: safety@iwap.co.uk.

Printed and bound by CPI Group (UK) Ltd, Croydon, CR0 4YY

12/05/2026

02108889-0001